The Modern Period

The Modern Period

Menstruation in Twentieth-Century America

LARA FREIDENFELDS

The Johns Hopkins University Press
Baltimore

© 2009 The Johns Hopkins University Press
All rights reserved. Published 2009
Printed in the United States of America on acid-free paper
2 4 6 8 9 7 5 3 1

The Johns Hopkins University Press
2715 North Charles Street
Baltimore, Maryland 21218-4363
www.press.jhu.edu

Library of Congress Cataloging-in-Publication Data

Freidenfelds, Lara, 1972–
The modern period : menstruation in twentieth-century America / Lara Freidenfelds.
p. ; cm.
Includes bibliographical references and index.
ISBN-13: 978-0-8018-9245-5 (hardcover : alk. paper)
ISBN-10: 0-8018-9245-7 (pbk. : alk. paper)
1. Menstruation—United States—History—20th century. 2. Feminine hygiene products—United States—History—20th century. I. Title.
[DNLM: 1. Menstrual Hygiene Products—history—United States. 2. History, 20th Century—United States. 3. Menstrual Hygiene Products—economics—United States. 4. Menstruation—psychology—United States. 5. Social Class—United States. WP 11 AA1 F899m 2009]
QP263.F74 2009
612.6'62—dc22   2008028728

A catalog record for this book is available from the British Library.

Special discounts are available for bulk purchases of this book. For more information, please contact Special Sales at 410-516-6936 or specialsales@press.jhu.edu.

The Johns Hopkins University Press uses environmentally friendly book materials, including recycled text paper that is composed of at least 30 percent post-consumer waste, whenever possible. All of our book papers are acid-free, and our jackets and covers are printed on paper with recycled content.

For Ome

CONTENTS

Introduction 1

1 Before "Modern" Menstrual Management
Keeping Secrets, Wearing Diapers, Avoiding Chills 13

2 The Modern Way to Talk about Menstruation
Education, The Scientific Narrative, and Public Discussion 38

3 The Modern Way to Behave while Menstruating
Changing Health Beliefs and Practices 74

4 The Modern Way to Manage Menstruation
Technology and Bodily Practices 120

5 Tampons
A Case Study in Controversy 170

Conclusion 193

Acknowledgments 201
Appendix: Interview Method 205
Notes 211
Essay on Sources 229
Index 237

The Modern Period

Introduction

From cloth "diapers" boiled on the stove and reused, to Kotex and Tampax. From shock at the sight of the first bleeding and an awkward explanation by an embarrassed mother to educational films and pamphlets in fifth-grade classrooms. From warnings to avoid swimming, over-exertion, and "mental shock" to reassurances that having one's period did not preclude any normal work or play. During the twentieth century, Americans adopted news ways of thinking about and managing menstruation, ways radically different from what many preceding generations had taken for granted.

Deciding to adopt these new modes of menstrual management was no small matter. It required Americans, during the late nineteenth and early twentieth centuries, to overturn practices and beliefs that Western women and men had valued and followed for centuries. They abandoned the conviction that regular menstruation was critical to women's general and reproductive health, and stopped worrying that mental or physical shock during menstruation could cause fatal injury. They gave up the thriftiness and self-sufficiency of homemade, cloth menstrual napkins in favor of the convenience, comfort, and expense of disposable pads. And they turned away from old concerns about keeping unmarried girls innocent about sexuality and reproduction in favor of early menstrual education. We should not take for granted the abandonment of these long-standing practices and beliefs but should recognize that they could only fall in the face of a particularly compelling confluence of new ideas and concerns.

The late nineteenth and early twentieth centuries saw dramatic changes in American patterns of childbearing, consumption, education, and work. Many more families came to value, and were able to attain, a lifestyle that included a small family, a high school education, a pink- or white-collar job, and comfort and status provided by a wide range of mass-produced goods.[1] This lifestyle entailed work, school and recreational situations that demanded a new level of attention to self-presentation and personal efficiency. To meet these demands, women adopted novel technology, knowledge, and practices to manage menstruation.

These new menstrual beliefs and practices were shaped by the Progressive values of the early twentieth century. It was an era of profound faith in the uplifting possibilities of scientific rationality, popular education, technological progress, and control of natural and social processes through scientific study and bureaucratic management. Reformers of all stripes believed that scientific and managerial approaches to business, government, and social affairs would produce a wealthier and more equitable society. Industrialization, begun in earnest after the Civil War and continuing into the early twentieth century, brought many more Americans into workplaces that were the targets of efficiency studies and labor reform. Men experienced the routinization of their work processes and the disciplining of their labor practices first, as they moved in large numbers out of artisanal and farm work and into factories. Women joined them in increasing numbers in the early decades of the twentieth century, many entering a burgeoning realm of clerical and sales work. Progressive enthusiasm had waned somewhat by the 1920s, evident in the decline of ambitious social reform, but many Progressive ways of thinking were absorbed more deeply into American culture.[2]

Americans applied Progressive values not only to their work and schooling but also to more intimate aspects of their lives. In thinking about menstruation, twentieth-century Americans envisioned the creation of a well-controlled body that would not leak, smell, hurt, cause anxiety, appear unfashionable, or lose efficiency (productive or reproductive) at inopportune moments. It would integrate specialized technologies seamlessly, so that the signs of menstruation and the technologies themselves would be invisible and undetectable to everyone, not least the woman using the technologies. Finally, a woman's menstruating body could be brought under control because it would be well studied and understood. Deviations requiring special attention would be anticipated and discussed in numerous settings—including school education programs, advertisements, magazine articles, and doctors' offices—and between mothers and daughters.

New approaches to menstruation were established publicly and privately in three interrelated areas. First, in cooperation with sex educators and corporate pamphlet writers, Americans embraced a scientific explanation of menstruation and advocated that it be taught to girls at menarche or earlier in order to allay anxieties about the onset of menstruation. Second, in conjunction with physical educators and industrial hygienists, women and men developed new health beliefs that led them to think that women not only could but should participate in all their normal activities all month and that these activities would not endanger their health or cause serious discomfort during menstruation. Third, collaborating with menstrual products manufacturers and advertisers, women enthusiastically adopted new technologies and techniques for better hiding menstrual blood and increasing comfort during menstruation. While these areas of reform were associated with separate sets of public advocates and institutions, they strongly reinforced each other, weaving together into a coherent vision of menstrual management.

In this book, this confluence of new ideas and practices is called the "modern period." This label is useful in two ways. First, it gives a single name to what might be otherwise perceived as disparate strands of change in education, technology, and health beliefs. This highlights the striking degree to which these strands of change wove together into a coherent whole, with an identifiable internal logic. Second, specifically labeling these changes as "modern" captures the spirit in which Americans made them and echoes the language popularly used to describe the move away from practices that "experts" and laypeople alike characterized as "old-fashioned."

This popular conception of a "modern" way to handle menstruation incorporated new social, cultural, and economic patterns that historians recognize as central to American modernity. Women channeled some of the most important modernizing forces of the turn of the twentieth century into their care for their menstruating bodies. They drew upon new ideas about universal education of children, and sex and hygiene education in particular, to give their daughters early explanations of menstruation in scientific terms. As consumer culture began to flourish, they looked to the consumer market to provide them with nationally marketed, branded, widely advertised, disposable menstrual pads. They took seriously the health advice from new experts in physical education and industrial hygiene, letting go of old concerns about protecting their bodies during menstruation so they could work and play all month. Women's bodies manifested in the most concrete and intimate terms what historians have understood to be major strands of American modernity as it was developing at the turn of the century.[3]

At the same time, the "modern period" as a popular vision of bodily modernity was not simply a happenstance convergence of these modernizing trends. Women drew upon and gave new shape to popular education, consumerism, and health advice in service of a particular kind of "modern" body, one that was ideally always efficient, predictable, and presentable, in line with Progressive values. They, and the educators, advertisers, and medical experts with whom they collaborated in envisioning and realizing new norms of menstrual management, explicitly named their innovations as "modern." This book explains the nuances, internal logic, and import of this popular conception of what it meant to manage menstruation in a "modern" way, as revealed in archival sources together with in-depth interviews with seventy-five American women and men of a range of ages and from several ethnic and regional groups. Documenting and analyzing this phenomenon of popular modernity is crucial to fully understanding not only the history of menstruation but also the history of American modernity.

In this popular conception, becoming modern carried with it large stakes: having a modern bodily self-presentation and self-control was perceived as critical to joining the middle class, and necessary to entering a growing women's job market. While women and men alike were under pressure to cultivate a "modern" self-presentation through daily bathing, tooth brushing, and other hygienic and grooming practices, women bore an extra burden. First, they were often the first in their families to hold a pink- or white-collar job and so were in the vanguard of their families' socially visible move into a perceived middle class. Second, their wages from clerical and sales jobs often was what gave their families the disposable income to support these new "modern" bodily practices and other accoutrements of a middle-class lifestyle. Finally, women were under particular scrutiny by employers as new entrants into a novel and growing space in the labor market.[4]

Becoming "modern" in this way was a significant part of what allowed the vast majority of Americans, by midcentury, to think of themselves as middle class, despite clearly evident variations in education, income, and prospects.[5] Commentators on the political Right have sometimes seen the expansion of a self-perceived middle class as a mark of American progress and superiority over other nations. Activists on the Left are often frustrated that Americans are hard to rally as "workers." Unlike in much of Europe, too many Americans who are "working class" by virtue of their economic status view themselves as "middle class" for political activists to be able to effectively rally them on behalf of what would appear to be their class interests. Americans have tended to believe in the American Dream, in pulling themselves up by their bootstraps, in individual betterment as a road to economic prosperity. Whether one is situated on the political Right and believes that more ardent bootstrapping is the obvious answer to American inequality or one is on the political Left and believes that working Americans are being hoodwinked into self-undermining alliances with the bourgeoisie, it is crucial to understand why and how the vast majority of Americans came to experience themselves as middle class during the last century.

Because we take modern menstrual technologies and practices so much for granted today, it might be tempting to ask, "Didn't women just adopt them because they were obviously better? Did wanting to be 'modern' really have anything to do with it?" While it is true that women did not generally say to themselves, "I want to become modern today," and go to the store or the library looking for the tools to do so, it is a mistake to assume that their adoption of modern technologies and practices was completely instrumental. The adoption of modern menstrual management required almost no knowledge or technologies that were radically new in the early twentieth century.

When it came to menstrual education, women always had viable explanations, even if they were not the same ones that would become available in the twentieth century, and they could potentially have shared them with their pre-menarcheal daughters. Prominent medical authorities had, in fact, been urging them to do so since at least the late eighteenth century. With regard to menstrual technology, disposable pads could have been made from cotton wadding, and given that early Kotex generally had to be disassembled and cut, re-stuffed or otherwise rearranged to make it fit an individual woman, the technological innovation was not actually a large step from earlier possibilities. Tampons were already in use for medical management of nonmenstrual vaginal bleeding or discharge. Finally, like education and technology, health beliefs could easily have shifted earlier. By the late eighteenth century, medical authorities had for the most part abandoned the idea that menstrual bleeding was a necessary discharge of "bad blood," which women had to carefully maintain in order to protect their health. Popular medical writers maintained an emphasis on the importance of regular menstruation, but they generally made little effort to support their claims with medical theory. American women could have adopted early menstrual education, disposable menstrual pads, and new beliefs about menstrual health long before the turn of the twentieth century, had it simply been a matter of using whatever technology and practices were most convenient and comfortable. Instead, it took the large-scale social changes of the early twentieth century to make these innovations, conceived of as part of "modern" bodily management, appealing and accessible on a mass scale.

Early twentieth-century social changes made menstrual knowledge and technology much more widely available, as they were promoted heavily by the new "experts" of the Progressive era. Sex education programs, menstrual product advertisements and promotion in drug stores, and free, readily available menstrual education pamphlets broadly distributed modern approaches to menstruation. Equally important were newly emphasized values of efficiency, convenience, and consistent, carefully monitored self-presentation, which in turn supported new roles for women, in school and in the workplace. Women adopted these "modern" values on a broad scale and, because of this, wanted a modern approach to thinking about and managing menstruation, and took advantage of the available information and technologies. By the 1950s, these new ways of thinking about and managing menstruation were highly elaborated and widely adopted, supported by a strong post–World War II cultural consensus around middle-class values.

The modern period was so established and taken for granted by the end of the 1950s that even the counterculture of the 1960s could not disrupt it. During the 1960s, the American counterculture defined itself partly in opposition to what it felt was a complacent, conformist, sexually repressed middle-class culture. However, when it came to menstruation, the innovations of the counterculture, including young women's use of the Pill, public discussion of sex and bodily functions, and a more matter-of-fact attitude toward the body, actually enhanced the status of the well-managed, modern body rather than undermining it. In the end, greater openness in discussing menstruation, an extension of the previous generations' Progressive values under the aegis of the sexual revolution, the 1960s counterculture, and the feminist movement allowed young women and men to expand and refine their parents' and grandparents' vision of modern menstrual management.

Young Americans in the final decades of the century took for granted most of the beliefs and practices their parents and grandparents had so carefully and self-consciously cultivated, and rarely consciously associated them with a particular class standing. Those who came of age in the 1980s and 1990s were even better at modern menstrual management than previous generations had been, using more refined, diverse, and broadly available technologies and information. Confident in their abilities to be "modern," they began to critique and modify the practices bequeathed by their forebears, stretching some of their predecessors' Progressive approaches to limits their parents and grandparents sometimes found shocking. Young women expanded acceptable spaces for acknowledging menstruation, at times demanding that the efforts they made to manage menstruation, and the annoyance menstruation sometimes caused, be recognized and supported. At the end of the century, young people took for granted the shaping of their intimate bodily experiences by Progressive values established decades earlier and the basis those values provided for a shared understanding of what it meant to belong to a broad middle class. From this base, young people began to critique and re-envision bodily modernity through their management of menstruation.

The creation of the modern period was a collaborative effort between a variety of newly minted "experts"—including sex educators, physical educators, advertisers, industrial hygienists, and manufacturers of disposable menstrual products— and ordinary Americans. The oral history interviews conducted for this book reveal that Americans deliberately rejected older models of menstrual belief and management. Women worked with public advocates of various stripes to create a new way to manage menstruation that they felt alleviated much of their previous shame, confusion, discomfort, and frustrating lack of control of their bodies and lives. While the twentieth century saw many new modes of social and individual attempts at bodily control that are rightly regarded as pernicious, such as forced eugenic sterilization of the "unfit" and high rates of anorexia among teenage girls,[6] new modes of menstrual management are not among these. Interviewees explained convincingly why they embraced most of the modern education, health beliefs, and technologies generated by various public advocates, and how these new modes of menstrual management generally allowed them to work, study, and play as efficiently all month as they and their bosses, teachers, friends, and partners hoped. Many also indicated that they experienced new, modern modes of menstrual management as integral to their transition into middle-class status, through their adoption of attitudes, knowledge, and bodily self-presentation that differed from their parents'. New expectations of bodily control were not necessarily oppressive; the history of menstruation reminds us that "modern" bodily management could be experienced as satisfying and potentially liberating.

This book's narrative is largely a story of agreement and shared belief, among women and men from different backgrounds, and also between women and the various public figures, such as manufacturers, educators, advertisers, and physicians, who advised and tried to influence them. This level of agreement and cooperation may seem surprising, given the scholarship of the last thirty years, which has so persuasively demonstrated profound differences among Americans on the basis of class, race, region, ethnicity, and other social distinctions. The research for this book began with the goal of looking for the ways in which Americans differed from each other and from the various public speakers who claimed to represent them, and instead uncovered an amazingly robust shared vision of "modern" menstrual management. Women and men from the three groups primarily interviewed—African Americans in the rural South, white New Englanders, and Chinese Americans in urban California, as well as an additional few interviewees from other backgrounds—were strikingly consistent in their desire to instantiate modern bodies for themselves, and they defined their desired results in nearly the same terms.

Not every aspect of modern menstrual management was uncontroversial. For example, tampons were eagerly adopted by some, while rejected as dangerous, too difficult to use, or ineffective by others. Many women who chose not to use tampons displayed a keen awareness of the class and ethnic differences their other, shared, menstrual practices had served to obscure. They pointed out that they were not entirely comfortable with the full range of modern bodily practices that they thought of as originating in middle-class, white, urban culture, even though in most ways they were interested in adopting such practices. Interviews with women from several different ethnic, regional, and class backgrounds are uniquely valuable in providing the evidence for two major claims of this book: first, that Americans from very different backgrounds did come to share and embody a vision of a "modern" body, which was an important basis for the twentieth-century emergence of a self-perceived middle class encompassing most of the population; and second, that modern menstrual management was rooted in Progressive ideals developed by the white, urban, well-educated middle class, and those roots were only gradually and incompletely forgotten.

The creation and adoption of modern knowledge, technology, and health beliefs happened amazingly quickly in historical time, given the robustness and longevity of the previous beliefs and practices that were replaced. In terms of an individual lifetime, however, the shift was prolonged, and many interviewees born before 1960 felt that they struggled to make the change over the course of their lives. Many observed in their children's and grandchildren's lives the continuation of the transformation that they had begun. While nearly all interviewees shared a vision of a modern way to manage their bodies, many adopted the associated practices piecemeal and over time.

The book's chapter organization, based around themes rather than time periods, brings into relief the sweep of change in each thematic area over the course of the century. It also highlights that these changes were gradual, fragmented, and occurred for different women at different times. The first chapter paints a picture of premodern ways of managing menstruation, as they were taken for granted at the turn of the twentieth century, and briefly examines their historical precedents. Chapters 2 though 5 then address four specific topics in the creation of modern menstrual management, tracing each from nineteenth-century precursors through the end of the twentieth century. Chapter 2 examines menstrual education and the scientific narrative of the menstrual cycle, as it was presented in written sex education materials and read, distributed, and discussed by those interviewed for this book. Chapter 3 looks at changing health beliefs about menstruation and how new beliefs supported and were encouraged by new expectations at work and school. Chapter 4 takes up new menstrual technology, as it was created, advertised, and used. Chapter 5 examines persistent controversy over the safety, efficacy, and sexual implications of tampons, and how these differences of opinion reflected interviewees' identities in relation to the white, urban, middle-class, Progressive roots of modern menstrual management.

In telling this history, it is important to listen to the voices of the women and men interviewed, who explained why they preferred "modern" menstrual management to "traditional" ways of handling it and how they collaborated with institutions and experts to create a new mode of menstrual management they generally found satisfactory. This book looks simultaneously at the stories of American women and men and the efforts of various experts and institutions, recognizing the powerful influence of experts but also the necessary cooperation of ordinary people as self-educators, consumers, patients, managers of their own bodies, and sometimes as self-conscious critics of experts and institutions. The bodily modernity they created may not always appear ideal from an early twenty-first-century feminist perspective. However, it is important to understand that women and men generally felt that the new approaches worked reasonably well for them. They worked in subtle ways to reshape this bodily modernity when they were dissatisfied, and they embraced it even though they were often aware of how its Progressive, white, urban, educated origins did not comfortably fit their own identities.

Historians of women have been suspicious of twentieth-century demands for bodily discipline and new modes of social control exercised through bodies, and rightly so. Many women have learned to obsessively monitor their weight, for example, yet despite the energy they put into dieting, they never feel good about their body size. In extreme form, women who strive for perfect control can become anorexic, making themselves seriously ill in their extreme efforts to control their bodies. Even in less extreme form, dieting can encourage women to be critical of themselves and spend the bulk of their energy on self-reform rather than improving their position in society.[7]

Feminist historians' justified critique of this type of self-destructive instantiation of a "modern" body ideal has been amplified by Foucauldian critiques of modernity in recent decades. Foucault argued that, beginning in the late eighteenth century, Western nations primarily disciplined their subjects not through imposing direct and violent force but through inculcating self-discipline in the context of institutions such as schools, factories, prisons, and hospitals. Foucault and his followers tend to regard this inculcation of self-discipline as a sneaky and insidious way for states, and those who hold power within them, to gain an even greater degree and more intimate level of control over their subjects.[8]

In addition, besides new regimens of self-discipline, the late nineteenth and early twentieth centuries saw many social movements for greater governmental and medical control of bodies, particularly those of marginal people. A eugenics movement aimed at improving the "quality" of the "race" (usually meaning white, Anglo-Saxon Protestants) led to the government-imposed sterilization of thousands of mentally disabled and "delinquent" Americans. Jim Crow laws in the South kept black bodies in "their place," and lynchings were ignored or encouraged by government officials. Reformers bent on stamping out prostitution often ended up promoting the imprisonment and poor treatment of the women they were ostensibly trying to protect. Temperance activists began by urging Americans, especially immigrant Irish, to have more self-control and to exercise moral pressure on each other to abstain, but eventually were able to pass Prohibition, at least temporarily.[9]

Given this confluence of attempts at social control through the direct exercise of government power and the encouragement of many modes of self-discipline, it is not surprising that scholars who have looked at the twentieth-century history of menstruation have tended to see many aspects of the modernization of menstrual practices as potentially ominous. Many worry that the large menstrual product manufacturers, and their advertising and educational materials, are usurping the rightful role of mothers in menstrual education and promoting the idea of menarche as a "hygienic crisis" rather than a socially and personally meaningful coming-of-age. Some extend their criticism to include pharmaceutical companies and physicians promoting extended-cycle oral contraceptives designed to reduce the frequency of a woman's periods and drugs to treat premenstrual syndrome. Most commentators are not critical of the modern ideal of widely available menstrual education books and pamphlets, but they think that these materials are too often produced by those with a vested interest in selling certain menstrual products. In addition, they believe that late twentieth-century educational materials, which ostensibly promote "openness" about menstruation, contradict themselves by emphasizing the various methods that can be used to most effectively hide bleeding, odor, and other evidence of menstruation.[10]

It is, indeed, important to note the efforts powerful institutions such as corporations and the medical profession have made to shape modern bodily practices to their own benefit. However, listening to the voices of ordinary women and men, and their explanations of why they found their collaborations with various experts and institutions satisfying and helpful, provides an important balance to conclusions historians have reached from archival sources alone. It is also an important step in explaining exactly why Americans, and especially American women, have so readily embraced new modes of bodily management that often seem against their self-interest. In fact, in some areas, including menstrual management, these new modes actually worked for them as promised most of the time and provided valuable rewards at school and work, and in self-perceived class mobility.[11]

In addition, looking at the stories of ordinary women and men next to published sources and institutional perspectives allows for a reinterpretation of what has appeared to other scholars to be the self-contradictory nature of modern ways to talk about menstruation. The kind of "openness" about menstruation that is part of modern menstrual management is very specific: it is openness in carefully circumscribed locations and constrained language, just what is needed to support the modern desire to make menstruation impinge as little as possible on people's lives, and no more. Women wanted to learn openly about menstruation in sex education class and buy menstrual products along with milk and bread in the grocery store because they wanted to manage menstruation more effectively, not because they had a general desire to be more open about their bodies. Menstrual education pamphlets that encouraged open discussion about menstruation in sex education classes and between mothers and daughters, and at the same time sold products on the basis of how well they hid menstrual blood, were perfectly consistent, understood through the logic of the modern period. Modern approaches to menstruation by and large are not contradictory; rather, they are so powerful and effective exactly because educational, commercial, and medical efforts consistently reinforced each other.

This book's conclusions rely on the juxtaposition of archival research with seventy-five in-depth oral history interviews. The appendix describes in more detail important aspects of the interviewing and interpretive methods used. While not every interviewee is quoted in the book, their stories all contributed to its conclusions. These interviews are an exciting and unique source. Nothing comparable exists in archival sources, in terms of the depth of individual information. Extant fragments of archival sources documenting individuals' experiences, such as an occasional sentence in a letter or notation in a diary, are almost exclusively from well-off white women. In the past thirty years, historians studying women, minorities, and the working class have spent much fruitful effort in uncovering the experiences of those who did not create written documents. Historians interested in reproduction and sexuality have likewise begun to document the history of experiences not usually recorded even by the literate. This book draws inspiration from both types of history, taking on the challenge of using a nontraditional historical source in order to expand the bounds of our understanding. The results demonstrate the value of this approach, allowing us to see how individual women's and men's choices about managing their bodies mattered at least as much as the commercial, educational, and medical viewpoints of the experts with whom they collaborated, in the creation of the modern period.

CHAPTER ONE

Before "Modern" Menstrual Management

Keeping Secrets, Wearing Diapers, Avoiding Chills

At the beginning of the twentieth century in the United States, "modern" approaches to menstrual management were just beginning to take shape, and most women still thought about and managed menstruation in ways that would have been familiar to many generations of their ancestors. They learned about menstruation only haphazardly and thought of it as shameful and embarrassing, something to hide from other women as well as men. They managed menstrual blood with homemade cloth pads, which most women washed and reused. They avoided swimming and sometimes even tub bathing because they worried that getting chilled during their periods could make them sick or halt the bleeding, a potential source of illness in its own right. And they considered sex during menstruation to be unhealthy, unpleasant, and even immoral. Men's knowledge about menstruation was gathered even more unsystematically then women's, mostly from other men, and they rarely talked about menstruation except when discussing their girlfriends or wives' sexual availability.

All of these patterns would soon be challenged by women and men collaborating with sex educators, physical educators, industrial health experts, menstrual products manufacturers, physicians, and other cultural authorities to create a "modern" vision of menstrual health and management. By the time the granddaughters of the oldest women and men interviewed for this book were coming of age, experiences of menstruation in the United States had changed dramatically. Interviewees took note of these changes over the course of their own lives and in the differences they saw between their own experiences and those of their daughters and granddaughters. They were able to describe vividly what they deemed "traditional" ways of managing menstruation because they contrasted so starkly with what they experienced as new, "modern" modes.

While these patterns of change were common in the lives of older interviewees, the transition between "traditional" and "modern" experiences of menstruation was not clean or abrupt. The half-century or so it took for women's ordinary, intimate experiences of their bodies to change drastically is remarkably short in historical time, especially given the weight of many centuries of significant continuity. In terms of individual women's lives, however, a half-century is a long time, and women did not adopt new bodily practices and beliefs all at once. Some women adopted "modern" practices selectively or gradually over the course of their lives. Some had become thoroughly modern by the 1920s or 1930s, while others did not make the transition until the 1950s or 1960s. Therefore, the stories of those born between the first years of the twentieth century through the early 1940s are relevant to understanding what were commonly considered "old" ways of managing menstruation, ways that had impressively deep historical roots.[1]

For women raised before the advent of "modern" menstrual education, learning about menstruation was a haphazard process. Many older women interviewees did not know about the existence of menstruation before menarche and were scared when it happened and felt hurt that no one had warned them. Christina Donvito, the daughter of a white Boston shipyard welder and a homemaker, recalled her experience getting her first period in the early 1950s.[2]

> I looked, I panicked, I was bleeding. Mom never told me nothing. And the kids, we never thought about stuff like that. Because I was a sports-minded girl. Played volleyball, basketball, tennis, did all these things at school. And one day, "Gee, I felt funny. What the heck's going on?" I looked. "Oh my God, I'm bleeding. Now I better tell Mom this, something's wrong." Walked into the kitchen, and it was just Mom and I, it was afternoon. Thank God I had come home from school—I was home when it happened. "Mom!" "What's the matter?" "I don't know how to tell you this," I says, "but I got *blood* all in my underwear." "Oh, honey!" and then she sat and taught me. I said, "Why didn't you tell me before? Why did you wait [for me] to get shocked—I thought it was something drastic." [Laughter.] I was probably twelve or thirteen. I said, "Why didn't you tell me?" She said, "You know, Christina, I just didn't know what to tell you. I just didn't know what to say to you."

Those who were aware of menstruation before menarche did not learn directly from their mothers. Many learned from friends at school. Roberta Cummings Brown, whose African American parents ran a small convenience store in the rural South, attended a small, segregated elementary school in the 1940s. There, older girls took pride in educating younger girls about methods of managing menstruation, the agonies of cramps, and the excitement of having boyfriends.

"They told you all the gory details," and at the same time, "they always sort of glorified it." Roberta's mother herself had been warned by her older sister, and it had never occurred to her that she might be expected to teacher her daughters about it. Roberta did not even tell her mother she had gotten her period until she decided she needed some help managing it. "It was trial and error as to how to take care of it. I finally had to tell, I finally did say to my mother—we had a store which sold sanitary napkins . . . she said, 'Oh, whenever you need any, just go get them.'" It took Roberta a while to even realize that her mother also menstruated. "She always hid her things. She never shared that this happened to her. And at the time, when I first began, I really didn't—probably my immaturity—I didn't think this was something that moms did."

Even those whose mothers did not hide menstruation so assiduously did not learn from them directly. Rachel Cohen, growing up in an economically struggling Jewish family in New York City in the 1930s, recalled that while her mother was not particularly forthcoming about menstruation, she did not specifically hide it from her daughter, any more than from her women friends.

> I took care of my little sister, I cooked meals, I did laundry, I cleaned the house, I went to the grocery, I did shopping, I did everything. And so I wasn't fenced off in a child's world. I was part of the adult world. So naturally, I was always hanging around. I was always there, and so anytime my mother talked to another woman about something, I heard it. And she didn't shoo me out or tell me to leave. She didn't try to keep anything from me. But she never voluntarily came to me and said, "This is going to happen." I guess she did with me what was done with her. You know, her mother didn't sit her down and give her this shining lecture about, "Now you're a woman." You just, you took care of things as they came along.

Although her mother did not have a notion of "childhood innocence" that needed to be maintained by withholding information about sexuality and menstruation, information only slipped out occasionally and casually, and reticence was the norm. For example, it took a long time for Rachel to figure out that when her mother said she was going out to "see somebody" and came home with wet hair, she was really going to *mikveh*, a ritual cleansing bath used by Orthodox Jewish women after each menstrual cycle. "She was too embarrassed to talk about it, and I never asked her." Still, because her mother had women friends who spent a lot of time at their home, Rachel felt she was able to collect all the information she needed. "As a girl, you hung around when the women sat and talked and you heard them talking and, you know, I just put it all together."

Given these different ways in which girls learned about menstruation—even in the same social milieu, and even within the same family—some individuals found out about menstruation before their first periods and others did not. Mary Hanson, the daughter of a New England factory engineer and a homemaker, learned enough from friends at school in the 1900s and early 1910s to be prepared when the moment came, but her younger sister did not. Mary found herself having to explain it on the spur of the moment to relieve her sister's shock and concern because their mother was hospitalized and was not there to do it. Rachel Cohen noted that living together in a tight space did not guarantee that sisters would know about each others' periods. "You always went to your own girlfriends. And [my sister] had a whole gaggle of girlfriends, and I'm sure they all discussed it. But I had nothing to do with her period."

Because of how they learned about menstruation, girls concluded that it was shameful and embarrassing. As Mary Hanson put it, "It was just almost like something you didn't talk about. That you should be ashamed of it, was what you almost thought sometimes. It was just a subject you didn't discuss." Since it was not a topic for open discussion, it was talked about in dark corners in secretive whispers instead. At school,

> Of course, they'd snicker and fool and talk, you know. They were embarrassed too, but they wanted to talk about it . . . we used to get our heads together and try to figure things out and talk. We thought that was wonderful to talk about it, because you couldn't talk anyplace else. So I think we learned a lot of things like that, maybe things we shouldn't have . . . It was always a big, dark secret. So that you used to love to get in corners and talk about it. Forbidden fruit.

In their awkward attempts to caution girls about their new vulnerability to pregnancy, mothers often added to the sense of confusion and shame. Christina Donvito's mother warned, " 'Well, you know, you're going through changes,' and this and that, and in those days, 'Stay away from boys. Don't let the boys touch you.' Of course, I was afraid to hold hands! I didn't question her, and that's all she said, was 'Don't let any boys come near you, now.' " Mothers' unwillingness to talk about sex openly could make the way they addressed menstruation baffling to their daughters. Roberta Cummings Brown, in contrasting her own attitudes toward menstruation with those of her mother, also linked the difficulty of talking about menstruation to embarrassment about sex. "I'm not prudish. I'm more open. And I look at it as a natural process, whereas she didn't. And she never told us where babies came from. She just didn't. That was a topic that she wouldn't discuss. And that goes back to her generation."

Boys' learning about menstruation was even more furtive and haphazard than was girls'. Unlike girls, of course, they did not reach a menarcheal deadline by which they would inevitably find out about it. Even in schools where sex education programs were instituted, they were rarely the target of formal education about menstruation. In addition, their mothers, sisters, and female friends took great pains to hide any evidence of menstruation from them. John Graves, growing up in an African American community in the rural South in the 1940s, believed that he first had some idea about the existence of menstruation because he "heard rumors at school." He also recalled asking his mother, a teacher, about blood he found in the toilet. His mother explained briefly that a visitor "was probably menstruating. She didn't go into too much detail . . . She just said, 'She shouldn't have been careless, not flushing the toilet properly.'" He learned a lot more, later on, from boys at school. "People would say, 'She's on her period,' you know, if a girl was excitable that day . . . some of the boys might mention that," though only to other boys, not directly to girls. He also learned about the pitfalls of having sex with a menstruating girl from his friends, who he guessed were largely making excuses for not being able to convince girls to have sex with them.

> The body's breaking down tissue, and stuff like that. I had, I guess that was the conversation, the boys used to mention that. It was not sanitary . . . And then fellows who would attempt to have sex with a girl said they had a bad experience, because the girl was so emotional, she might scratch him, or whatever. [Uncomfortable laughter.] If they pressed their way with her, they ran into some real difficulty sometimes.

John pointed out that he would never question a partner if she told him she was menstruating, but he acknowledged that some men and boys might, or might at least brag that they had tried. Concerns about girls' and women's moods and sexuality dominated male discussions of menstruation throughout much of the century, and for much of this time, boys and men learned about menstruation primarily from their male peers.

Among those interviewed for this book, this pattern continued through the 1960s. As Mike Ozols, the son of white New England family farmers, remembered, boys still learned about menstruation primarily from other boys at his rural high school in the 1950s. "Just the guys talking about it . . . Talk about the women having the rag on, and stuff like that. They couldn't go out, they were feeling sick—stuff like that. We'd kind of joke around. We didn't really understand what it was . . . Every once in a while they'd leak a little bit, and they'd get all embarrassed—we'd be laughing at them." While he and his friends liked to joke around at the girls' expense, they did not tease the girls directly. "Back in our time, it was really kind of bashful. Wasn't as open as it is now."

Among the women interviewed, communicating about menstruation with boys and men—whether they were fathers, brothers, sons, friends, or schoolmates—was rare and generally considered undesirable; the only exception was certain kinds of communication between husbands and wives. Mary Hanson's friends reinforced this sensibility. "They used to say, 'Don't go around the boys. Be careful. Don't let them know that your period is going on. Don't discuss it in front of boys.' That was a warning we got. As if we would discuss it! You wouldn't. You'd only discuss it with a girlfriend. It'd have to be a close girlfriend, too."

Women reported taking extra pains to hide evidence of menstruation from their fathers, almost imagining their fathers did not know that they menstruated. Roberta Cummings Brown explained that she and her sister hid it from their father because they felt "embarrassed." "At the time, I looked at it not as a good, not as a normal thing. It was just my immaturity, and my thinking at the time was men should not notice. Not, you know, why—it never occurred to me that he already knew. These boxes [of Kotex from the family store] were disappearing, first of all. [Laughter.] But anyway, he never said anything." Christina Donvito made a special effort to help both her sister and her mother hide menstruation from her father, telling them to keep their supplies with hers, in a special box in her closet. She was particularly careful about how she disposed of pads, for the same reason.

> Nothing ever went down the toilet. No, no. Oh, God, no. [Laughter.] Because that would have been embarrassing—tell Daddy the toilet was clogged, and those things come flying out! You've got to think of these things, you know! [Hearty laughter.] You've got to think of these things, right? That's why you never put this stuff down there. Yikes!

Women also hid menstruation from their sons. Mike Ozols reported that his Eastern European–born mother, Mara Ozols, never revealed any evidence that she menstruated. He only guessed she did after he found out that girls at school had periods. "With four boys growing up, she never kept stuff like that [pads or tampons] around . . . Maybe she was embarrassed to show it to us, or something. You know, start asking questions, it gets embarrassing. She wasn't much one to talk about anything. That, and sex—anything like that." When interviewed, Mara objected to ads she had seen for menstrual products on television precisely because they might force a mother to talk about menstruation with her sons. "I thought disgusting. Nasty. I say, all kids watching. 'Oh, Mommy, what's that?' Little boy come, 'what's that for?' I thought that was wrong. OK, you got in store, you put in some magazine or something. On TV I don't agree with. That's somehow overboard. To me. Because I am old-fashioned."

Communication between teenage girls and boys growing up in the 1940s and 1950s was less rare among the African Americans interviewed than among other groups. Menstruation was not an open topic, but boys sometimes teased girls directly about being "on the rag," and while girls disliked this teasing, they expected it. Roberta Cummings Brown explained what this teasing was like.

> Boys in particular would look for this [menstrual] belt, and then they would say horrible things, like, "Oh, she's wearing the rag" . . . They would just say it behind, you know, like boys do, behind the back . . . You could hear them . . . And it doesn't have to be *you*, you know. You would know that if it ever happened to you, and sometimes it did, you sort of ignored it . . . [Also] boys always would go, "Mmm, something smells bad in here." Or, "Something smells fishy." You know, they would use that expression. And the last thing I wanted to be was to smell fishy . . . they'd pass by and go, "Mmm, something smells fishy." And even if they didn't, that was just their way of trying to be funny.

John Graves did not remember talking in front of the girls in his school, but his conversations with other boys may not have been as private as he thought.

African American interviewees were also more likely to actually mention menstruation between boyfriends or girlfriends in the context of teenage sexual interactions, while women from other groups were more likely to try to avoid that conversation altogether by breaking dates when they got their periods. As Ida Smithson, the daughter of tobacco farmers, recalled, girls in her school in the early 1930s whispered about other girls who had sex during menstruation. John, growing up in the same community in the 1930s, said that on a date, "you might hear a girl, if you tried to get fresh, if she liked you she might tell you . . . 'Oh, you can't feel me there, I'm on my period,' or something like that." In contrast, when Christina Donvito was asked whether her high school "steady" ever knew she had her period, she replied, "No way! No way, no. [Laughter.] Not at all." When asked if she thought he ever wondered why she would not go out once a month, she answered, "He probably did, but that was OK. It wasn't for discussion."

The one exception to the avoidance of mixed-gender conversations about menstruation was when women were communicating with their husbands about their sexual availability. All of the married women interviewed, of all ages, said decisively that they would tell their husbands they had their periods if he made sexual overtures. Many said that their husbands often knew anyway, simply from being physically close, and they just assumed that it was reasonable for a husband to know these things.

The experiences reported by the oldest interviewees, of learning haphazardly and in whispers about menstruation, of feeling that it was embarrassing and shameful, and of considering it important to keep it secret, especially from men, had a long history. All of these patterns can be documented, though some more thoroughly than others, at least as far back as the Middle Ages in Western culture.[3]

As historian Monica Green has shown, surviving sources mentioning menstruation in medieval Europe are almost all medical texts, and they generally refer to menstruation as "the flowers," or sometimes as "monthlies" or "catamenia." The few mentions of popular perceptions, however, indicate an attitude of secrecy and shame. A translation of one of the most important midwifery texts, the *Trotula*, gave alternatives to popular names for women's reproductive organs and for menstruation because "women are ashamed to name these things." The translator suggested that menstruation be called "les fleurs" as a substitute for "the 'customary' name *les malades secrettes*." A number of other medieval medical texts on diseases of women also suggested that women might conceal health problems associated with genitals or menstruation from male doctors because they were ashamed to reveal or talk about these things. In the later Middle Ages, a new genre of writings about women's bodies and generation, intended for a male audience not necessarily made up of doctors who treated women, was commonly labeled "Secrets of Women," and gave a particularly negative picture of menstruation as dirty, dangerous, and shameful. Green also notes that the almost complete silence around menstruation in nonmedical works, even in texts containing daring and explicit sexual joking, suggests that menstruation was a taboo topic, particularly in mixed company.[4]

Menstruation was kept secret, especially from men, in a world in which a great deal of information about women's reproduction was held and exchanged within the community of married women. Throughout the early Middle Ages, births appear to have been attended almost exclusively by female midwives and married women neighbors and relatives. In the late Middle Ages, wealthy women began to consult learned male doctors if they had problems with pregnancy or the postpartum period, but midwives generally continued to handle the birth process itself, and male doctors were expected to refrain from handling female clients' genitals. Nonelite women continued to receive reproductive care from midwives and from each other, and in America, this remained true through the nineteenth century. Women supported each other in postpartum care, and some wealthy women paid poorer women to wet nurse their babies. The labor of pregnancy, birth, breastfeeding, and child care was expected to be exclusively the province of women, and this work absorbed much of the fertile years of most women.[5]

While this reproductive labor was conducted within households, in close proximity to male family members, the oral and tacit knowledge surrounding it was the property of the community of women. Religious and legal doctrine demanded that wives be subordinate and meekly obedient to husbands. At the same time, women's daily reproductive work largely happened outside the direct purview of the men who had legal dominion over them. In its hiddenness, especially from men, menstruation was treated as part of this women's domain of reproductive knowledge and labor. Through the eighteenth century, in Europe and the American colonies, most women continued to bear children throughout their fertile years, and reproductive labor continued at the center of women's daily lives.[6]

There is no direct evidence before the late eighteenth century about whether girls were told about menstruation before menarche, but indirect evidence suggests that they were unlikely to have been routinely informed. While there is evidence of a strong sense of female community, in which women shared intimate knowledge of the body with each other, "women's secrets" were not shared with unmarried women. Those unlucky or imprudent enough to get pregnant out of wedlock found themselves shut out of regular channels for information about birth control, abortion, pregnancy, and birth.[7] Only married women attended each others' births. Unmarried girls were not a part of women's networks of bodily knowledge, so there is no good reason to assume they were regularly informed about menstruation.

By the late eighteenth century, health advice writers were explicitly telling mothers that they needed to do a better job of educating their daughters before menarche, implying that secret keeping was the long-standing and expected pattern. In 1769, William Buchan wrote of menarche,

> The greatest care is now necessary, as the future health and happiness of the female depends, in a great measure, upon her conduct at this period. It is the duty of mothers, and those who are intrusted [sic] with the education of girls, to instruct them early in the conduct and management of themselves at this critical period of their lives. False modesty, inattention, and ignorance of what is beneficial or hurtful

at this time, are the sources of many diseases and misfortunes in life, which a few sensible lessons from an experienced matron might have prevented. Nor is care less necessary in the subsequent returns of this discharge. Taking improper food, violent affections of the mind, or catching cold at this period, is often sufficient to ruin the health, or to render the female ever after incapable of procreation.[8]

Buchan's advice was echoed by health writers throughout the nineteenth century and into the early twentieth century.

Victorian values likely exaggerated these experiences of secrecy and ignorance for middle-class girls and women during the nineteenth century. Middle-class white women were supposed to be pure, delicate, and "passionless." At the same time that physicians urged women to look to doctors for reproductive care, they praised them for their modest blushes during examinations, and interpreted their discomfort not in terms of age-old traditions but in relation to the cultivated delicacy of Victorian sensibilities. Romanticized as morally superior (though legally and intellectually inferior) to men, women were charged with providing a peaceful sanctuary in homes newly dedicated to maternal domesticity and consumption. As women found ways to drastically reduce the number of children they bore, middle-class mothers were expected to engage in more emotionally intense mothering and provide longer-term maternal protection and moral education for their children. During this time of growing belief in childhood innocence, women were supposed to maintain the purity of their children, shielding them from outside influences, including the whispers of friends who might educate them prematurely about sex.[9]

African American women, coping with a very different set of circumstances, also tried to protect their daughters by suppressing information about puberty and sexuality. According to historian Marie Jenkins Schwartz, enslaved women in the antebellum years "restricted information about sexuality and menstruation to mature women in the hope that young women would remain naïve or modest and not grow up too fast." At a time when enslaved women's sexual activity and reproduction were often coerced in a variety of ways, and African Americans were assumed by whites to be sexually aggressive "Jezebels," enslaved mothers hoped that withholding knowledge would be a means to "shield girls from entering into marriage too soon and from the sexual abuse that sometimes followed a young woman's entry into puberty."[10] White, middle-class Victorian women and enslaved African Americans alike could call upon a long-standing vernacular tradition of sexual knowledge and attitudes, still strong in the nineteenth century, to support their convictions that the best way to shepherd young women into adolescence was to withhold information about menstruation until menarcheal bleeding necessitated a conversation about it.[11]

———ⱭⱭⱭ———

Experiences of shame and ignorance were not the only long-standing patterns echoed in the stories of the oldest women and men interviewed for this book; their health beliefs related to menstruation also had historical precedent. When Mary Hanson found herself unexpectedly in charge of telling her younger sister about menstruation, she reluctantly took on this duty because her sister was frightened and did not know what to do. "I was a little bit scared myself. But I did. I had to." She showed her sister how to use Kotex and "explained to her that she would have it every month. And that's another way she was getting rid of her bad blood. That's what they used to tell us."

Belief that menstruation was a way to get rid of "bad blood" was common during the early decades of the twentieth century. Related to this belief was the idea that a woman could get sick from retained menstrual blood if she didn't have regular periods or a period was suppressed. Emil Novak, in his 1935 guide for women, rather condescendingly noted that "the belief that absence of the flow is injurious is obviously a survival of the old superstition that the purpose of menstruation is to rid the body of harmful substances. Practically no competent authority holds any such view now, and yet many of the laity still cherish these old beliefs."[12] Historian Leslie Reagan notes that in the early decades of the twentieth century, women still often described early abortions as "bring[ing] my courses on," or being "put straight." Regular menstruation was necessary to health, and a missed period could indicate a health problem as easily as it could indicate pregnancy.[13]

While older interviewees did not alter their daily habits much during menstruation, the one specific health practice most of them maintained was related to this idea of menstruation as a necessary purging of the body of "bad blood." Mary Hanson and Samantha Fried, the daughter of an Eastern European Jewish immigrant and a Southern farmer, had been told that they could catch pneumonia if they swam during menstruation; the body was vulnerable while it was more open than usual to the outside. Mary also learned from her parents to avoid tub baths during menstruation. "You could sponge yourself off, but you had to be careful. Because you could get pneumonia or something else." Liza O'Malley, born in the Boston area in the mid-1930s to a machinist father and homemaker mother, avoided swimming because she had heard that it could stop her period, and that it was important that once it started, it continued. Most other older interviewees also avoided swimming, based on parents' advice, but they did not learn the basis for that advice. Unlike women a generation younger, though, they understood the reason to be primarily a concern for health rather than a concern for the cleanliness of the water.

These health beliefs were consistent in Western culture, with some variations, from ancient Greek times through the early twentieth century. Medical authorities from Hippocrates onward embraced versions of a "plethora" model of menstruation: women's bodies produced more blood than they were able to consume in their regular activities, so menstruation was necessary to rid the body of this excess blood. This conception of menstruation fit within a humoral understanding of the human body, in which flows of various fluids (including blood) in and out of, as well as within, the body were necessary to preserve health. Unbalanced, excessive, or stagnating flows caused disease, and medical treatment was aimed at bringing bodily flows back into balance.[14] While menstruation was generally acknowledged to be related to reproduction, in that menstrual blood was retained during pregnancy in order to provide matter to form the fetus, its importance to general health was often more strongly emphasized.[15] While women were considered to be colder and moister than men, and therefore in need of more regular purging, the plethora model of the body applied to both sexes, and in certain circumstances men as well as women could "menstruate."[16]

There were some important variations on the plethora model of menstruation. Many authors believed that the menstrual flow removed bodily impurities. Pliny the Elder, in the first century C.E., compiled a long list of common beliefs about the noxious, poisonous nature of menstrual blood.[17] These beliefs were strongly echoed in the secrets of women literature of the late Middle Ages.[18] The alternate point of view was that the blood was simply excessive and prone to putrefying in the body if it was not released regularly. Soranus expressed this view in second-century Rome; by the seventeenth century, it dominated the medical literature.[19]

Given this belief that menstrual blood, whether noxious or simply overabundant, needed to be released in order to maintain health, medical writers were most concerned about menstruation when it appeared to be stopped up. To a lesser extent, they also expressed concern when it flowed overly abundantly or too frequently and threatened to weaken the body. Medical treatment was based on balancing flows in the body. If the blood was thought to be stagnating in the body, doctors worked to evoke flows in various ways, and they also tried to strengthen the body so that the blood could flow properly. Treatments were aimed at the condition of a particular body, as in all humoral medicine, rather than at a particular disease, as in most modern medicine. Women used a huge variety of emmenagogues (medicines to promote menstrual flow), either when their periods had ceased and they did not believe themselves to be pregnant, or to provoke the initial flow in adolescents who had reached their mid to late teens without menstruating.

Stagnation of menstrual fluid was thought to be most commonly caused by shocks to the system, especially at the time when menstruation was expected or had already begun. Systemic shocks included exposure to cold and chilling, mental shocks such as sudden fright, profound grief, or even extreme joy, and the consumption of particularly hard-to-digest or overly sour or spicy foods. Exposure to cold created a double jeopardy because it could stagnate the menstrual flow and menstruating women were also considered to be particularly susceptible to bad colds, which could develop into chronic, fatal conditions such as consumption (tuberculosis). Overexertion or overwork was not a prominent concern, except in cases of excessive bleeding; then women were advised to lie down and stay quiet during their periods to keep from aggravating the condition. Medical advice books mentioned treatments for painful menstruation, but this was a minor concern next to absent menstruation.[20]

Through at least the Middle Ages, and significantly longer among the lower classes, almost all women must have frequently missed periods. Hard physical labor, lack of protection from cold temperatures in winter, inadequate nourishment, chronic illness, frequent pregnancies, and breastfeeding are all likely to cause amenorrhea, and these were common conditions for most women. Without pregnancy tests, it could be difficult or impossible to tell the difference between early pregnancy and the onset of illness. With the combination of medical concern about menstrual regularity and the reality of frequent irregularity, recipes for emmenagogues abounded and were distributed widely in medical and popular texts, as well as diaries and family recipe books.[21]

In the mid-eighteenth century, medical writers began to question various aspects of the plethora model. Some doubted whether menstruation was really caused by systemic plethora; they proposed a model of local plethora of the reproductive organs instead.[22] Humoralism was on the wane, at least in elite medical circles. Medical writers began to see the body as composed of discreet organs with specific functions and secretions, rather than as a continuous system of flows, and menstruation began to be seen as a specific secretion of the uterus.[23] Some began to suggest that suppressed menstruation should be seen as a sign of illness affecting the function of the uterus, rather than as the cause of that illness.[24] During the nineteenth century, medical researchers proposed new explanations for menstruation that challenged the plethora model further. One widely adopted theory held that menstruation was the equivalent of heat or rut in animals; menstrual bleeding was seen as analogous to estrus bleeding in female dogs.[25]

However, despite all these challenges to the plethora model of menstruation in the medical literature, it persisted in whole or in part in many medical advice books for laypeople throughout the nineteenth century. For example, John Gunn's extremely popular *Domestic Medicine,* first printed in 1830, indicated that prolonged suppression of menstruation could lead to the spitting of blood and other symptoms; in that case, "immediate attention should be paid or consumption will take place."[26] In his 1874 publication *Plain Home Talk,* Edward Foote reflected an understanding of menstruation based on the plethora model when he described how suppressed menstrual blood tended to flow vicariously from the lungs, nostrils, mouth, eyes, stomach, or rectum at monthly intervals if it did not find its natural outlet. He then declared that "menstrual derangements should never be neglected, for in all cases, excepting suppression by pregnancy, they lead to other diseases which are liable to prove troublesome, and perhaps fatal."[27]

Even those who no longer explained the purpose of menstruation in terms of the plethora model inherited concerns about amenorrhea from previous centuries of medical belief. Nearly every nineteenth-century health manual warned that one exposure to shock or cold resulting in suppressed menstruation could cause health problems for life. In 1847, physician and birth control advocate Frederick Hollick explained menstruation in terms consistent with the local plethora and "heat" theories: "The ripening of the ovum causes a local excitement, and congestion, in the ovary and womb, which increases till the period when it is thrown off, and then the accumulated fluid is discharged, the excitement subsides, and a new development commences."[28] He did not appear to believe that menstrual blood was part of a systemic overload of blood or a carrier of bodily impurities. Nonetheless, he declared that "there is scarcely a single disease that [the menstrual cycle's] derangement will not either cause, or at least seriously aggravate. It is therefore *vitally important* to attend to this matter, *particularly in young persons approaching puberty!* A little care at that time, properly bestowed, may prevent years of disease and suffering, if not untimely *death!*"[29]

Hollick was more alarmist than average, but his advice reflected an ironic trend in the nineteenth-century health literature. At the same time that the long-standing plethora theory that explained the need for regular menstruation was fading, physicians were becoming increasingly convinced that menstruating women were delicate and vulnerable to disease. They particularly began to emphasize the vulnerability of girls during puberty. They forcefully scolded mothers to inform their daughters about the impending onset of menstruation because they feared that an uninformed girl might have suppressed menstruation from sheer fright at the event or from attempts to staunch the bleeding with applications of ice or cold, damp cloths. Like Hollick, they warned that suppressing menstruation one time at puberty could have lifelong consequences. The stridency of their warnings is striking next to the lack of any real attempt to explain why, outside the outmoded plethora model, suppressed menstruation would be so dangerous. It appears that while many popular health writers had begun to follow elite researchers in questioning old ideas, the plethora model—in combination with general Victorian ideas about female delicacy—still had a powerful influence on their perceptions of menstruation. The belief that regular menstrual bleeding was crucial to women's health, and that menstrual suppression through shock or cold was dangerous, would not be seriously challenged until well into the twentieth century.

Another long-standing health belief was that sex during menstruation was potentially risky for both partners, in addition to being immoral. During the Middle Ages, religious texts condemned sexual intercourse during menstruation as a sin, while medical texts warned that it was dangerous to the male partner and that a child conceived during menstruation was likely to be leprous or epileptic.[30] By the early twentieth century, scientific and educational experts had mostly put aside the idea that menstrual blood was inherently dangerous, "bad" blood, labeling that idea a superstitious taboo dating from Greek and Judeo-Christian antiquity. Still, they were equivocal about the health implications of sex during menstruation. Many medical researchers and sex educators were willing to refrain from completely condemning the practice of sex during menstruation, largely because they believed that women experienced greater sexual desire at this time, consistent with the idea that menstruation was equivalent to "heat" or rut in animals. According to these experts, sex during menstruation might be necessary for "marital harmony."[31] Even those who disapproved of sex during menstruation thought that it happened with some frequency because women had such strong sexual desire at that time of the month. According to Charles Malchow, writing in 1923, "It is not so much the want of desire on the part of the woman as it is the consciousness of being unfit, or the fear of contaminating or causing disgust, that prevents the proper mood for intercourse. Not seldom passionate women will tolerate or invite relations when there is some show of menstrual discharge and afterwards plead ignorance of their condition and excuse their conduct with profuse apologies when evidence of the bloody discharge becomes exposed."[32] In accordance with the ideas of famed sex researcher Havelock Ellis, Malchow believed that women would demonstrate greater desire for sex at the time of menstruation, if only there were not cultural taboos in place. While "women of refinement" would not consider sex during menstruation, according to Malchow, prostitutes often pursued it, and because it enhanced their menstrual flow, they were less likely to become pregnant.

Malchow was not the only author who thought that sex during menstruation could have specific health implications. Hugh Northcote, in a 1907 sex education book aimed specifically at Christians, did not deny the possibility that women experienced greater desire for sex during menstruation but claimed that the aversion to sex during menstruation was not simply a taboo imposed by the Christian religion. "This instinctive aversion must rather be regarded as part of nature's design in the sex life of humanity. It is a sexual safeguard to women in a condition of catabolism."[33] Other versions of this idea pointed to the swollen state of the women's reproductive organs, engorged with blood, and expressed concern that they could be sore and tender after intercourse, or even damaged. There was also concern that the cervix was open more than usual, leaving the uterus more vulnerable to infection.[34] Some older women interviewees expressed similar ideas. Ida Smithson agreed with these health concerns, regarding the postpartum period and menstrual periods as requiring the same respect. "Well, I thought your body needs healing, to go back to its—just think about it, everything was out of shape, and all of this. And you want it to go back to the shape it's supposed to be." Christina Donvito felt similarly. "That was just kind of a time that my body was going through an adjustment, a clean-out feeling to me," and therefore it seemed unhealthy to have intercourse.

Sex education literature also expressed concern for men's health, as did some older male interviewees. It was noted in the literature that bacteria was carried in menstrual blood and that gonorrhea could be passed to a man more easily if he had sex with an infected woman during her period. However, the literature usually dismissed concerns about sex during menstruation causing urethritis.[35]

Given all of these health concerns, sex educators were divided over whether it was reasonable for a couple to have sex during menstruation, respecting the woman's presumed peak of sexual desire, or whether it really was best to abstain. Those who thought it should be allowed, generally emphasized that it should be done by those who did not have any reproductive health problems, and then only in moderation and with an emphasis on cleanliness for the woman.

The oldest women and men interviewed for this book had never actually considered having sex during menstruation. They were the most likely to think of menstrual blood as "dirty" and assumed that sex during menstruation would be repulsive to men. As Mary Hanson commented, "Well, didn't they tell us it was our bad blood? Why would a man want to have anything to do with it?" She never discussed it directly with her husband; they both assumed that sex during menstruation was a bad idea. She never considered the possibility of having sex during menstruation, though she had heard recently that young people now do, and she was rather put off by the thought of it. Mara Ozols, who heard people on a talk show say they had "good sex" during menstruation, was also quite taken aback by the idea. This disapproval could go so far as moral objection to sex during menstruation; Ida Smithson said offhand that women and men who choose to have sex during menstruation "don't have no scruples."

Older women and men interviewees agreed that it was important for men to respect women's wishes to abstain during menstruation. Unlike the feminists of the next generation, who would criticize men for regarding women's bodies as disgusting or unfit for contact during menstruation, women of this generation criticized men who pushed them to have sex during menstruation. Ida Smithson considered men who said, "You are my wife, you're my girlfriend, I tell you what to do," to be unscrupulous. Liza O'Malley had only had sex during menstruation when she was coerced by her husband, and she considered the behavior abusive. "I think the only time when I ever had it was when my husband was drinking and I couldn't say no. So there was a big turn off on it, then . . . He would be drinking, and nothing else mattered but satisfying himself." Men agreed that if a woman said she could not have sex because she was on her period, there should be no more questions asked. John Graves recalled teenage petting on dates in the early 1950s; sometimes a girl would refuse to be touched and explain that she was on her period. When asked if he could tell if that was sometimes just a convenient excuse, he said he could not be sure. "I never pressed; I never pressed them for information. I just accepted their word, that's all." He followed the same practice in his marriage.

Through the early part of the twentieth century, self-appointed experts on sex and lay women and men agreed overall that sex during menstruation was not a good idea. They expressed moral and medical concerns, and tended to think that abstaining was an appropriate way to show respect for the aesthetic, moral, and health concerns of male and female sexual partners alike.

In addition to patterns of holding traditional health beliefs, feeling ignorance and shame, and practicing secrecy surrounding menstruation, the oldest women interviewed for this book recalled using traditional technologies for managing the practicalities of bleeding. Before Kotex made its debut in 1921, only a portion of the established middle and upper classes created their own disposable pads; many middle-class and most poorer women used cloth.[36] With the advent of Kotex, many more women adopted disposable pads, but in poorer communities, the use of cloth persisted to some degree at least through the 1940s.

Practices for making cloth pads varied, but all were part of what Susan Strasser has described as part of the ancient tradition of *"bricolage,"*[37] the creation of something new from the spare parts of something old, or what a Kotex educational pamphlet would later deride as "old-fashioned, inadequate, make-shift ways," compared to "practical, efficient, up-to-date, Kotex."[38] Ida Smithson explained, "We used, just, old sheets, old things that you had around the house, and things like that." She and Jane Cummings, born in the rural South to African American factory workers and raised largely by her brother after her parents died young, did not remember anyone showing her or her friends how to cut or fold the cloth, or how to attach it to their clothes, but Jane dismissed that as an issue. "That wasn't hard to figure out. Of course, you didn't have anything else, anyway." It seemed obvious to her because she did not see any other choice of materials, she was likely used to improvising from leftovers, and also perhaps because it was similar to how she had seen baby diapers made. Mary Hanson explained, "We didn't have Kotex. We had diapers. You don't remember wearing diapers. You don't know. As a baby you had them. But they weren't very comfortable . . . you'd have to shape it, fold it over, just as you put on a baby." Mary remembered pinning the cloth to a special belt, made for that purpose, while Jane pinned hers to a string. Ida recalled following her mother's instructions to "go find a cloth, and pin it inside your pants." At a time when menstrual technology was for the most part improvised from household scraps, the technology women created was much less standardized than it would become with widespread use of Kotex.

Once the cloth "diapers" were soaked with menstrual blood, some women washed them and saved them for next time, while others threw them away. Their practices were on the cusp between an older culture of saving and reusing, and a newer culture of disposability.[39] Washing out cloths was quite a chore and created notable publicity problems. As Mary explained, "Of course, you had to soak them and wash them and cleanse them and blah, blah, blah, and hang them up so no one would see them. [Short laugh.]" For her, this meant hanging them in the bathroom overnight and putting them away quickly in the morning, since the normal drying line was outside in plain sight, and everyone in her household used the bathroom during the day. Ida, who threw hers away, commented with a note of incredulity on those who did not. "I remember people in my neighborhood who used to wash it—they used to wash those cloths. And save them for the next time! It was something to think about." Bodily habits marked class differences, and hanging one's menstrual cloths on the clothes line demonstrated that one was too poor to follow the more "hygienic" and "modern" practice of throwing away used menstrual pads. Within a household, it was even harder to keep secret. Not only did clean cloths have to be hung somewhere to dry, but washing them thoroughly could involve boiling them on the stove, creating a smell that Dorothy Joyce, the daughter of white New England factory workers, remembered vividly from a childhood friend's house.

Given the difficulty of keeping menstrual blood and technology out of sight when washing cloth "diapers," it is easy to see the appeal of throwing them away. Ida and Jane used outdoor privies during the time that they used cloth pads, so disposal was simpler than it would be at any other time in the century: cloths could simply be thrown into the pit with the rest of the waste. Jane opined that while this did use up a fair amount of cloth, periods only came twelve times a year, so it really was not bad compared to what would have been required for baby diapers.

Using cloth had a couple of conveniences that were only evident in comparison to the newer technology of disposable pads. First, a cloth "diaper," folded in several layers, was thick and absorbent enough that it did not need to be changed during the day. Since women of this generation reported that they did not carry pocketbooks to school with them, it was easiest to use a menstrual pad that would last until they returned home.[40] Ida speculated jokingly on how she could possibly have carried a cloth with her to school.

> Look, we didn't have book bags then. We carried our books in our arms. And no specific place to put your books or anything; you put them under your desk. You know, you had a chair, and in the chair, underneath, it had a rack where you put your books. The only way you could carry one was to carry it in your lunch bag! [Hearty laughter.] And I don't think you want to be carrying something like that in your lunch bag! . . . Unless you carried it in your bosom, now. You know, we could always find somewhere! [Hearty laughter.] Well, if you had large bosoms, you could hide it up there!

Second, for those who threw away used pads, storage of pads required no secrecy, since the cloth did not really become menstrual technology per se until it was used.

However, all of the women interviewees judged that the conveniences were far outweighed by the annoyances of wearing cloth pads. Mary explained that "diapers" were as hard to hide when she was wearing them as when she was washing them.

> Those old rags that you put on, and they scrunch up in the middle, it would be all messed in the middle, and the ends would be . . . you wore a belt, and you could tell, back there, you know, in the front it stuck out, the safety pin, you had a little bulge out there, and all this, unless you wear loose clothing . . . Of course, you didn't wear slacks and things. If you ever did, they would've shown because they were so awkward. But you wore dresses, so that problem [didn't arise], but today you never would do it.

Some aspects of the material culture in which these women lived, including outdoor privies and loose dresses, made the use of cloth menstrual pads less burdensome than it could have been, but did not persuade the women interviewed for this book to continue to use cloth once they had access to Kotex.

In the early part of the twentieth century, women improvised menstrual pads from cloth scraps or sometimes purchased disposable materials, and those who did not yet subscribe to the new culture of disposability dealt with the annoyance and embarrassment of surreptitiously washing bloody "diapers" each month. Kotex would have obvious appeal when it appeared on the market in 1921, given the discomfort and inconvenience of cloth pads, and growing expectations that women would work and attend school with their usual efficiency all month. By the 1940s, clear class lines would be drawn between the mass of women who enjoyed the middle-class comfort of Kotex and the truly poor women who could not afford them and continued to use cloth pads.

All of these centuries- or millennia-old patterns regarding menstruation—a sense of ignorance, shame, and secrecy; concern for the elimination of "bad" or superfluous blood through regular menstruation; a belief that sex during menstruation was potentially immoral and dangerous; and the use of cloth pads to manage menstrual blood—provided the social and cultural background for interviewees born in the West. Many of the older women and men interviewed for this book, however, came from China, so their medical concerns and moral beliefs about menstruation were based in a different medical and social system historically.[41] In practice, however, Chinese interviewees' beliefs meshed easily with those of their Western counterparts, even when the theoretical reasoning behind them was different.

Older Chinese interviewees consistently recalled a similar sense of shame and secrecy surrounding menstruation, like that surrounding sexuality in any form, and were unlikely to have been informed by their mothers about menstruation before menarche. Immigrants from China also shared similar concerns about regular menstruation and worried about contact with cold water during menstruation. Since the early medieval period, Chinese medical theorists wrote extensively of women's need for regular menstruation, and provided a wide range of treatments to regularize women's periods. As in traditional humoral Western medicine, within traditional Chinese medicine menstruation was considered part of the necessary regular flows of the body, and suppressed menstruation indicated a stoppage in the body, which would endanger health. Emphasis on the importance of regular menstruation, and its centrality to good overall health, increased over the medieval period in China and remained prominent in traditional Chinese medical approaches through the twentieth century. During the twentieth century, Western medicine gained ground in China and existed alongside and sometimes intermeshed with traditional Chinese medicine. In the late twentieth century, Chinese health education books based on Western medicine advised women to avoid cold, damp, and overexertion during their periods, consistent with traditional Chinese medical belief as well as common advice in Western health texts through the mid-1960s. Sex during menstruation was also considered potentially damaging to women's and men's health. Chinese American immigrant interviewees abstained during menstruation; two specified that their husbands thought it was dangerous to the man's health.[42]

Traditional menstrual technologies, too, were similar but not identical to those used in Europe and North America. Immigrants from China used homemade pads much longer than their American-born counterparts. All those interviewed for this book made their own menstrual pads as teenagers, out of paper that was bought in sheets and cut up for toilet paper as well. While they lived in China, a few eventually switched to pre-made pads, likely imported from Europe or the United States. Allison Zhou, the daughter of an internationally active academic and a health worker, remembered her father coming home from international business trips with mysterious packages for her older sister. "I also remember the boxes. Not too many of them, because I guess they were expensive, and so my mom also always said, 'This is for Gloria,' and I don't know what she gets that we don't get." In a 1985 study in Taiwan, Charlotte Furth and Ch'en Shu-Yueh's interviewees reported that poorer women could not afford the paper to make pads, never mind the imported pads, so they used cloth, and earlier in the century, poor rural women often simply allowed their clothes to soak up the blood.[43]

Even relatively comfortable women living in Taiwan and Hong Kong used different grades of materials for their menstrual pads, depending on their families' resources. Laura Hwang, the daughter of an army general and a homemaker, thought about it in terms of generational change. She explained that when her mother was menstruating, "they have some kind of tissue paper, made out of the grass, you know, very rough. I feel sorry for them. Like maybe that's why my mother doesn't like to even mention it. And by the time, we used some kind of paper, it's much better than what she had." Isabel Mao, daughter of a government official and just a few years older than Laura, could not afford the nicer paper, and she described the consequences of using the grass paper. "We were using very hard material, you know, like the papers, very rough. And even cut you, you bleed. Because I remember when I ride bicycle, when I was in college, I have to ride bicycle, we moved far. And even I was bleeding because of the rubbing. So that's how we grew up. I never had any comfortable cotton to wear." Bonnie Kwan, daughter of a businessman and a homemaker, explained how this kind of paper had to be used in order to make it manageable.

> At night, every woman, they just use the hand to soften it, you make it wrinkly and soften it . . . make it look like a shape like a rectangle. And then they fold it this way. So whenever, they could use these for all day long if they not soaking through. Whenever after they went to bathroom, they just turned this around [chuckle] and then I have the cleaner side towards your body.

Some women supplemented this kind of pad with a rubber lining, but as Phoebe Yu, the daughter of a government official and a homemaker, recalled, the extra protection had its disadvantages. "Especially during the wintertime when it got very cold, the plastic got very hard, and it could be very scratchy on the skin, and that could be hurting! . . . In Taiwan, during the wintertime, there is no heating. [Laughter.]" Recounting their troubles with uncomfortable menstrual technology, immigrants from China seemed almost astounded to remember how much harder their daily lives were in China than in the United States.

Under ordinary circumstances, women born in the early decades of the twentieth century could manage their menstruation in a way that was generally satisfactory or at least sustainable. These generations of women, however, went through two world wars, during which circumstances were not at all ordinary, at least for some. Several interviewees explained how the displacements of World War II made menstrual management much more difficult, and having to deal with menstruation in turn made displacement more difficult. Growing up in Eastern Europe, Mara Ozols was used to making her own pads, cutting up old clothes or other scraps and stitching them into an appropriate shape on a sewing machine, and then washing and bleaching them after each use. But as a war refugee in Germany, even the relatively basic components of material culture that she needed to take care of her periods in this way were not available.

> Was big fun in Germany, during war. There was not pins, and no pads. Nothing . . . in Germany was the worst thing. You couldn't even—oh, God. Couldn't wash and couldn't make. I don't know how I survived. [Laughter] . . . We don't have nothing . . . I think maybe I cut some sheet or something and I made, because we don't get nothing. That was terrible, and I hated . . . And sometimes no place to wash, and you save in some bag and then soak them and make sure there's soap. There was no Clorox to bleach them, even . . . Somewhere you find water, you wash. Make sure you wash, also make sure you don't lose, because you couldn't get any new ones.

It was not only immigrants who dealt with the indignities of wartime displacement and the difficulties of menstrual management away from accustomed environments. Californian Jill Okada Sun, daughter of a small businessman who lost his business permanently as a result of internment during World War II, remembered distinctly her experiences in a Japanese internment camp in the United States.

> My mother was prepared because we were of that age, when we went to camp. She had a roll of cotton, you know, it used to come wrapped in paper . . . She set aside some fine cotton, soft cotton material. She said, "We don't know where we're going. We don't know what's there." And so she said you might have to make your own pads. Which we did at first. I don't know what the other people did. Because Kotex was not readily available. We didn't have a PX to go to right away. That was something that came

later. So my mother had foresight and I marvel at that sometimes, with all the things she had to worry about . . . getting rid of everything and you know, just reducing everything to two suitcases per person to have stuff, the cotton and the material, and, of course, the needle and thread to make your own. [Laughter.] You cut it, cut the strip of cotton, lay one, and then another soft piece on the top and then you would just baste it. That's what we did, or what my sister did, anyway, perhaps the first year or the first three-quarters of a year.

Jill was lucky that her mother had planned so carefully, and was willing to sacrifice much of the little space for their belongings for the comfort of her daughters, but no amount of planning could solve the second problem at the camp—the lack of privacy in the bathroom. Jill knew more than she intended about other women's menstruation for architectural reasons.

We didn't talk about it much, between our friends, or you know. Just that . . . well, the bathroom situation in camp was not as private as one would have liked. So, you know, you knew when someone started a period. What we did was, at first there weren't any partitions in the latrine. Then, each block had a manager. Each block had maybe 250 people . . . And then the manager, you took your concerns and complaints and everything else to the manager. And one of the first things that most women requested was some partitions. So then it was up to the block manager to get lumber. You know, somehow get lumber, and we did get partitions. But, you know, to the side, but not in front, so that if there were two rows, you would try to go to the last one if you had your period. That's about as private as it was. So that's how we dealt with it, just discreetly. [Laughs a little.]

While Jill's difficulties were much less extreme than Mara's, they are in some ways more outrageous. While everyone around Mara was suffering the same deprivation, Americans who were not of Japanese descent did not even deal with Kotex shortages during the war, never mind the kind of humiliation Jill faced in the internment camps. For both women, though, having to deal with menstruation in an unfamiliar, relatively barren material environment was difficult, uncomfortable, and embarrassing, and neither older practices nor the newer ones developed as part of modern menstrual management worked.

In sum, women born in the early decades of the twentieth century were likely to think about and manage their menstrual periods in ways that linked them to many generations of their forbears. Those who subscribed to the long-standing ways of thinking about menstruation which preceded the broadly adopted modern innovations of the early twentieth century learned about menstruation haphazardly, more from friends than from mothers; learned that menstruation was somehow connected to the dangers of sexuality and pregnancy, but that its purpose was to cleanse the body of bad blood; managed their periods with improvised, and often reused, cloth pads; abstained from sex during menstruation on health, moral, and aesthetic grounds; and avoided talking to men about menstruation, leaving them to learn what little they could from each other. They remembered menstruation as a mystery, an annoyance, and a source of embarrassment and shame. New public advocates in American culture—sex educators, physical educators, and advertisers—would soon disrupt these old patterns, working with women to try to create new ways of talking about and managing menstruation, aimed at dispelling the mysteries, reducing the annoyances, combating the embarrassment, and turning menstruation into something that could be more easily mentioned but generally did not need to be because it was so well controlled. Public advocates would collaborate with women to rework ideas and practices surrounding menstruation. Women would adopt new ideas and practices because they alleviated long-standing difficulties and because they were necessary to creating a new, "modern," and ultimately middle-class way to manage their bodies, in the context of new expectations at work, school, and play.

CHAPTER TWO

The Modern Way to Talk about Menstruation

Education, the Scientific Narrative, and Public Discussion

A 1934 book called *Your Sex Questions Answered: What Every Modern Mother and Father Should Know* told parents that before a girl's first period, "she should understand that menstruation is not getting rid of impurities, but that the blood and other material which would nourish a potential child has to be discarded when not used for procreation; that Nature has established menstruation for that purpose, and that it is a bodily function about which she should have neither fear nor shame."[1] At the beginning of the twentieth century, this understanding of menstruation was alien to most Americans, who were indeed likely to view menstruation as a way of ridding the body of "bad blood," and to feel ashamed and embarrassed about it. By the turn of the twenty-first century, the book's reassurances sounded quaint, almost ridiculous. By then, the vast majority of Americans took for granted the "modern" perspective provided in the sex education book. During the twentieth century, Americans made a radical transition in perspective, coming to embrace a scientific explanation of menstruation and believing that it should be shared with all girls before they got their first periods.

The stories of two interviewees, Ida Smithson and Samantha Fried, demonstrate how women from very different backgrounds shared this trajectory and experienced their own attempts at self-education in the context of the development of "modern" menstrual education. Ida Smithson was born in the late 1910s into a large, tobacco-growing, African American family in the rural South. Her parents' house was "a small home for all of these people to live in," with "beds everywhere" and no plumbing, but "when you don't know any better, and everyone around you are living about the same, it's comfortable." She had six older sisters, but still, her first period came as a total surprise.

> When I learned, I was thirteen. I had no knowledge of what was happening to my body at all, and it was frightening. My mother hadn't told me *any-*

thing, not *anything*. Because, I don't know why. And my sisters didn't ever tell me anything. It was like a hush, hush thing. And so I remember being in school, and I had this accident, I thought, and I came home crying, telling my mother. She didn't put her arms around me; she didn't do anything. She just said, "Well, go find a cloth, and pin it inside your pants." And that's all she told me! And she didn't tell me to change it; she didn't tell me anything. She did not tell me it was going to be from now on! She did not say anything, whatsoever. So, quite naturally, there's always some wise children, in school, and I would ask them.

Ida relied on the whispers of her school friends for more information, and while her school had health programs that taught children how to brush their teeth and use soap, it did not have menstrual or sex education. At the height of the Depression, Ida went on to graduate as valedictorian of her class, the first in her family to finish high school. She had thoughts of college but could not afford it, and instead worked as a beautician and a housekeeper, and ran a family-owned store with her husband.

Later, after she married, Ida learned more about menstruation and reproduction from her doctor. "He was just a family doctor, and I learned more from him (bless his soul, he passed), I could go in his office and sit down, I could ask him very, very personal things. He didn't care how many people were sitting out waiting. He took his books down, and went through, with these type of things like that. He was very helpful." Ida also bought her own books

about my body, and all of that, that's where I learned, with pictures, with colored pictures. It told me about how the babies, and how they were in the body, and all of that . . . After I had my first children, I started to buy them, because I wanted to be educated myself. I would leave the books out, too. I did not hide them, or anything. They were for them to read *if they wanted* to read them.

Ida also offered more information to her daughters specifically about menstruation. She contrasted her approach with her mother's.

I wouldn't want my grandchild, no one, to go through what I went through being D-U-M-B, dumb. [Laughter.] Even to my last daughter, at this time, they were still using Kotex, and I gave her the little book to read. I talked to her, you know, there used to be a little book you read, and you talked to her, you know—coming of age. I talked to her, so I told her if she saw something, bleeding or something, I told her that her little belt was there, with everything ready. And so she came to me, and told me, "Well Mama, it has started." You see? And I kept confidence in her. And then once you keep confidence in them, it's far different than if they've got to learn it from somebody else . . . [Schoolmates] aren't honest with you. They'll laugh at you, tell you untruths. And that's so sad."

Like many women born in the first half of the twentieth century, Ida made a conscious decision to think about and talk about menstruation differently than her mother had. She embraced written, scientific sources and used scientific explanations presented in a matter-of-fact manner as a much-preferred substitute for the confusion and shame that surrounded her own experience of menarche. She felt that learning and talking about menstruation using medical and scientific terms and ideas improved her own feelings about her experience and gave her resources for better parenting. These themes run through Samantha Fried's story as well, despite tremendous differences in the two women's backgrounds.

Samantha Fried was born in the early 1930s in a small Southern town. When her parents divorced in the early 1940s, she and her sister and mother moved to New York City, where she discovered that her mother spoke Yiddish and had strong ties to the immigrant orthodox Jewish community in which she had been raised. When Samantha got her first period,

I didn't know what was—it was frightening. My mother also, I don't know who explained to her, but my mother, surprisingly enough, had very peculiar ideas herself, so it was a frightening kind of thing. She told me what I needed to do at that time, what I need to take care of myself, and her other admonition—"and don't let any boys come near you!" What did that mean? I had no idea what that meant. Why did that enter into this at all? [Laughter.] So that, the first part of it was frightening, but being in New York, we lived in a place that had a little library. The grammar school that I went to, even though it was a public school, it was an experimental school. So we went to the library a lot to do research kind of things. So I knew the librarian well enough that I asked her to help me find some books. So when my mother scared the hell out of me, I decided that I wanted to find out as much as I could what this meant. [Laughter.] So it was the idea of finding out what was happening to my body.

Before going to the library, she had approached her cousin, who gave her even more reason to seek out some alternative sources of information.

I have a vague memory of talking to my cousin, who scared me even more. I had a cousin who was about the same age. The understanding of what was happening, it was something to do with—it was something terrible and sinister that happened, and this was a kind of way of relieving the body of some kind of, almost, illness, and that it was necessary that this happened, so that you wouldn't be sick. The whole thing just sounded very weird, and very scary. To find out that this was not what it was at all was pretty interesting, and pretty reassuring. The street information, as such, in its particulars, was more frightening. And you must realize, at that time nobody talked about these things. So that you couldn't just go ask somebody. But it was OK to go to the library and get some books about the physiology of what was going on, so that you could find out.

She learned from the library books

that the menstrual cycle was readying, it was the process by which the womb was readied for reproduction, readied to receive a baby, and that when that didn't happen, that was why you had the menstrual cycle. And that was something that made sense. This is what the body is doing, so it has nothing to do with all those things you're spinning around it; it's just the body's way of doing what it has to do at that specific time.

In her first year of high school, Samantha had a class at her all-girls school which gave information about menstruation and pregnancy as part of teaching girls about their future roles as wives and mothers.

I think they listed it under some silly name like Home Economics, but it really had to do with the role women played. And it really wasn't controversial because it just had to do with being aware of your own body, and raising children, and taking care of a house. But it was all based on information—as opposed to learning habits, you learn the information, and the choices you have in handling the situations as they come up.

When Samantha's much-younger sister turned to her for advice,

I said that Mother had not had anybody to explain it to her, so she had a few ideas that were not quite as precise as they might be, and that I had had an opportunity to have other information, and that she didn't have to be afraid. And by then, I could draw her diagrams of what was going on in her body, to give her some idea of what this meant, and that it was nothing bad that was happening, and that this was all the preparation of her body to be a mature woman.

Later, Samantha raised her own daughter in a liberal, urban community in the Boston area in the 1960s and 1970s, among other college-educated parents.

It was a different time and place, there was a lot of talk about the whole growing into womanhood. Everything was open, so that we talked a lot before. She knew what to expect; there was no big problem with it. I answered questions. We also had a wonderful doctor, who talked to her. I think it was just a different time and place, where everything was open, and there was no fear . . . There was an openness of information, not only between us, but just, in society as a whole.

Samantha Fried and Ida Smithson, women with different backgrounds and life courses, born about twenty years apart, shared a common experience of rejecting their mothers' attitudes toward menstruation and turning to published, scientific explanations for a more authoritative and positive account of their bodies' functions. Americans from many different backgrounds participated in creating a new, "modern" way of understanding and teaching their children about menstruation.

It was easy for Ida and Samantha to find books that could teach them about menstruation. Those books were part of a long lineage of popular literature intended to educate lay people about their bodies, reproduction, and medical care. Long before the mass of American women became determined to teach themselves and their daughters about menstruation, the published resources for doing so were becoming available. Popular health manuals have existed in one form or another almost as long as books have been printed. An educated woman of the seventeenth century might have referred to Nicholas Culpepper's *Compleat and Experienced Midwife,* which contained a concise explanation of menstruation as the release of a plethora of blood that needed to be regularly expelled from a woman's body to maintain her health, an explanation that had dominated medical and popular thought for many centuries. It also gave detailed descriptions of the symptoms of and remedies for menstrual suppression, irregularity, and flooding.[2] Many American households of the Revolutionary era owned a copy of

William Buchan's *Domestic Medicine*, which did not explicitly explain the purpose of menstruation but addressed concerns and gave recommendations similar to Culpepper's.[3] Culpepper and Buchan assumed that women would seek medical advice if their periods were irregular in any way and encouraged them to think of regular, moderate menstrual bleeding as crucial to women's general health.

While advice books had long been available, popular books proliferated in the nineteenth century. Broad social changes made books of all kinds more accessible: more people could read, and technologies for printing and distributing books became cheaper over the course of the century. The demand for medical books in particular increased as various sectarian medical systems gained popularity beginning in the 1820s. Many people believed that they could successfully treat themselves with the help of new books about herbal medicine, homeopathy, or more mainstream medical theories, rather than solely on the basis of tradition or family books of recipes for remedies handed down through the generations.[4] For example, John Gunn's *Domestic Medicine* (1830), one of the most popular home guides, gave laypeople detailed instructions in the use of mainstream medicine's heroic therapies, including use of opium and laudanum and bloodletting, for regulating menstruation, though without theorizing about the purpose of monthly periods.[5]

Just as people treated themselves using information from popular medical books when they did not have the money to pay a doctor, or doubted the efficacy or safety of mainstream medicine, they also dosed themselves with patent medicines. The Lydia E. Pinkham Medicine Company, which sold Lydia Pinkham's Vegetable Compound, one of the best-selling patent drugs for women's reproductive problems, distributed several pamphlets educating women about their bodies, as did the makers of Mattson Syringes and Dr. Ralph's Pills. All of these patent medicine makers gave some information about menstruation that reflected current medical understanding, but, of course, they shaped their narratives to make their remedies appear to be the solution to all menstrual problems.[6]

Another genre of books, advocating knowledge of reproduction and contraception, came out of still more major social shifts of the nineteenth century. Religion and religious institutions gradually declined as a source of authority in American culture, and science and scientists came into ascendancy. Popular interest in science, and scientific explanations of the human body and its workings, grew. At the same time, Americans grew less fatalistic about fertility and childbearing and began to take effective action to control the size of their families. Freethinkers, who rejected the authority of religion and sought new philosophical bases for social and individual decision-making, authored volumes that combined philosophical justification for contraception with scientific explanations of reproduction and fertility control.[7]

While books focused on sex and contraception had to be purchased more surreptitiously than popular medical manuals, both were widely available. This continued to be true even when the 1873 Comstock Laws forbidding distribution of contraceptive information as pornographic suppressed a certain amount of published information about reproduction. Smaller, less reputable publishers and distributors filled the gap.[8] Public lectures as well as books provided scientific information about the body and reproduction to some nineteenth-century Americans. Itinerant lecturers gave talks about anatomy and physiology in small towns across the country, while Ladies' Physiological Societies had regular meetings in larger cities.[9]

With all of these books and lectures available, and with so many copies of the most popular books distributed and sold, it would be easy to assume that scientific explanations of the menstrual cycle were common knowledge, and that parents would have this information to share with their children. Older interviewees' stories, however, make it clear that such was not the case. It seemed to them that their parents did not have any scientific understanding of menstruation; if they did, they felt too ill at ease to try to explain it to their children, and they certainly did not share any books with them. There are several possible reasons for this disparity between the broad availability and distribution of sources of information about menstruation by the end of the nineteenth century, and interviewees' experiences at the beginning of the twentieth century.

First, there was no broad consensus among medical researchers about the purpose of menstruation or the relationship of menstruation to ovulation until the 1930s. Ancient ideas recently had been rejected by most medical writers, but a satisfactory alternative had yet to take their place. Eighteenth-century medical writers had begun to link menstruation more tightly with reproduction and the amenorrhea of pregnancy, and by the end of that century were speculating that the menstrual plethora was really a plethora only of the female reproductive organs, not of the entire circulatory system.[10]

Research on ovulation in animals in the early nineteenth century led scientists to link menstruation not only to gestation but also to a monthly cycle of fertility. After 1843, when spontaneous ovulation, rather than coitus-induced ovulation, was demonstrated to occur in dogs and theorized to occur in women, most scientists in Europe and the United States came to believe that menstruation was linked to a monthly occurrence of ovulation.[11] Between 1843 and the early 1930s, however, there was little agreement about just how menstruation was related to ovulation, and what the purpose of menstruation could be. Many researchers explained menstruation as the equivalent of "heat" in animals; they theorized that menstruation was like the vaginal bleeding of dogs in estrus, signaling that they were fertile. Most believed that ovulation occurred simultaneously with menstruation. Some speculated that the beginning of menstrual bleeding created a surface to which it was easiest for a fertilized ovum to attach, and if it did, menstrual bleeding would immediately cease; others theorized that menstruation cleaned out the uterus, leaving a new, healthy surface for the implantation of the fertilized ovum.[12]

The scientific landscape was confusing and contradictory. In addition, the most popular theory, postulating that menstruation was equivalent to "heat" or rut in animals, seemed repulsive and counterintuitive to many who responded to these theories in writing, and probably to lay readers as well.[13] Given longstanding taboos against sex during menstruation, it seemed a strange suggestion that this was when women would be most fertile and most sexually interested. Finally, since these theories, unlike the plethora theory, did not appear to have much bearing on medical treatment, freethinkers could not particularly tie their scientific theories to their medical recommendations. The result was that while freethinkers shared their enthusiasm for science, they did not provide a straightforward, consensus view that readers could easily understand or confirm from other sources.

While authors of popular texts were tremendously enthusiastic about science, they practiced it with somewhat less precision than the researchers they referenced. Instead, they drew on whichever theories they found most compelling, sometimes creating a mish-mash of conflicting theories. Versions of these theories appeared in popular books about health and reproduction. John Harvey Kellogg, in his tremendously popular *Plain Facts for Old and Young* (1882), as well as other texts such as *Ladies' Guide in Health and Disease* (1891), said that menstruation was exactly parallel to "oestrus," "heat," or "rut" in animals, giving a drawn-out comparison in the *Ladies' Guide*. He urged mothers to tell their daughters about menstruation, but given how squeamish mothers appear to have been about mentioning menstruation at all, it seems unlikely that many sat their daughters down and told them that when they had their periods, they were like rutting animals.[14] Others gave more palatable explanations; Edward Foote called menstruation "Nature's wash-day," saying that the womb needed to be cleansed of unused ova and other detritus deposited there by the ovaries.[15] In his *The Diseases of Woman* (1847), Frederick Hollick combined the local plethora theory with a comparison of menstruation to the sexual cycle in animals. A nineteenth-century reader who perused more than one of these texts might have absorbed these authors' emphasis on scientific explanation but would almost certainly have emerged feeling even more confused about the purpose of menstruation than she or he had been before reading.

While popular literature about the body, reproduction, and sexuality blossomed robustly in the nineteenth century, the production and purchase of the literature itself was not enough to guarantee that it was understood, embraced, or shared with the younger generation in a family. The path Ida Smithson took, from shock at her own menarche, to a determination to do better with her own children, to consulting the available literature and sharing it with her children, does not appear to have been at all obvious. It took new, "modern" attitudes, the development of scientific consensus, and new venues of distribution for a scientific understanding of menstruation to truly become widely disseminated. Even during the twentieth century, it continued to be a gradual process, in which multiple generations of women experienced external and internal struggles in rejecting traditional attitudes and practices and adopting modern ones.

The Progressive Movement of the late nineteenth and early twentieth centuries provided a crucial base for the changes interviewees experienced. In menstrual education, as in other aspects of menstrual management, the ideas and practices interviewees described as "modern" were very much of a Progressive flavor. Progressive reformers believed in the value of scientific explanation and the rational management of sexuality. Most importantly, they strove to share these values with all Americans. In the nineteenth century, an emerging middle class had adopted "good manners" and hygienic practices such as frequent bathing and changing of clothes in order to distinguish themselves from the lower classes. In contrast, the Progressives were interested in extending these middle-class values and mores to everyone, in an attempt to "Americanize" the masses of immigrants and migrants from rural areas entering U.S. cities. They created public and private programs, wrote books, and petitioned legislatures to regulate work and housing to create sanitary conditions. An important Progressive legislative victory was mandatory public schooling, which was broadened to include instruction in hygiene, manners, and sometimes even sex education.[16]

Progressive reform efforts reached not only immigrants but also many native-born poor blacks. Middle-class African American reformers targeted women and men like Ida Smithson's parents, raised on small family farms in the rural South in the 1900s and 1910s, and raising their own children in the same circumstances in the 1920s and 1930s, as well as poor blacks who had migrated to cities. They believed that African Americans could "uplift" themselves and disprove racist stereotypes by adopting norms of "respectability" from middle-class white culture. They encouraged African Americans to repudiate the distinctive African American culture that had developed under slavery and assimilate to white American culture in all of their daily habits of work, speech, dress, and religious practice, as well as personal hygiene.[17]

Reformers came largely from the ranks of "new women" bursting onto the public scene from all directions. These middle-class "new women" were college-educated, often forwent marriage and motherhood in favor of career, were concerned about women's political, legal, and social well-being, and were dedicated to improving it. Their efforts were not entirely novel: they built on the reform efforts of Victorian middle-class women, who had joined voluntary organizations aimed at bringing middle-class "domestic" values of piety, sexual purity, and sobriety to the wider community. However, in their assertive participation in the public sphere and not infrequent rejection of motherhood and domesticity, they represented a break with the past. For white women, this meant escaping the gilded cage of the middle-class home. For black women, it meant having the luck and wherewithal to gain a formal education—often being the first in their families to do so—and escaping a life of domestic service or agricultural labor. African American and white middle-class reformers shared the public scene with young, unmarried, white, working-class new women, who pioneered independent urban living for unmarried women and participated in mixed-sex entertainment in public venues such as dance halls and amusement parks. Whether middle-class reformers or young, unmarried workers, new women in the Progressive era began publicly rethinking American sex roles and sexuality.[18]

In the 1890s, Progressive reformers began to publish books aimed specifically at educating teenagers about their bodies and sexuality. Earlier texts had urged mothers to warn their pubescent daughters about the impending onset of menstruation, but the texts themselves and the explicit information about anatomy and physiology they contained were intended for adults. Like Ida later on, married people would obtain these books, perhaps after having a child, perhaps before beginning their sexual relationship. Even books for teenagers were not intended for pre-menarcheal girls, who were assumed to be far too young to handle information about sexuality; it would take until the 1940s for widespread pre-menarcheal menstrual education to begin. But these books were an important precursor, the beginnings of educators writing for young people about sexuality directly, rather than through their generally reticent parents.[19]

Progressive sex education in schools was also aimed at teenagers, not prepubescents, and was largely focused on scaring teens into abstinence by showing graphic examples of the ravages of sexually transmitted diseases. But it also often included information about anatomy and again was an important precursor for later school sex education programs.[20] The Progressive emphasis on scientific explanation and popular education about sexuality outlived the Progressive era and was enhanced by the increasing sexual frankness of the 1920s. By 1927, a survey of high schools revealed that of the several thousand schools responding, 45 percent of them taught some sort of sex education. Of those that offered programs integrated into biology or physical education classes, only a quarter discussed menstruation,[21] probably because most of the students were well past menarche, but this fraction still represents a significant beginning to menstrual education in public schools. Teaching guides for sex education, including menstrual education, also proliferated during the early decades of the twentieth century, giving teachers a better chance of figuring out how to approach a delicate and seldom-discussed subject.[22] The Girl Scouts of America developed teaching aids, as well, and by 1921 required girls to learn about the physiology of menstruation to earn the Health Winner Badge.[23]

Public libraries slowly began to put sex education books on their shelves, although doing so was controversial. In 1934, in the pages of the highly subscribed, leading-edge trade magazine *Wilson Library Bulletin*, librarians debated the propriety of stocking sex education books and sharing them with teenagers. While a number of those who wrote in were not comfortable with libraries providing sex education, still, the majority of respondents as well as the *Bulletin*'s editorial page believed that libraries had a duty to make sex education information available, even to teenagers.[24] Samantha Fried's public grammar school, while on the progressive edge in its willingness to stock sex education books in the 1940s, could easily have been following the lead of the many public libraries by then willing to circulate sex education books and pamphlets.

Among the most important sources of information about menstruation were the pamphlets created by menstrual product manufacturers, especially Kimberly-Clark. By 1923, just two years after introducing Kotex onto the national market, Kimberly-Clark began publishing pamphlets aimed at girls and women. Its first efforts were scattered, self-contradictory, and short-lived, but Kimberly-Clark persisted. Out of a commercially driven desire to be considered a friendly authority to which all women and girls would turn for health and hygiene advice and menstrual products, Kimberly-Clark produced reams of mildly progressive educational materials. A few of them read like extended Kotex ads, but many played down the commercial angle, hoping to gain customer loyalty through public service rather than direct advertising. These booklets, designed to be inoffensive to the vast majority of women, may not have been revolutionary in the frankness of their content, but they contributed tremendously to a revolution in the availability of information about menstruation.

Pamphlets for women provided a great deal of straightforward scientific explanation and health advice, which women could use themselves or use as inspiration for advising their daughters. During its first twenty-five years of creating pamphlets, Kimberly-Clark's commissioned authors managed to write booklets for girls which provided enough information about menstruation to seem useful to parents and daughters but not so much detail about sex or reproduction that parents were reluctant to share them. Although these menstrual education pamphlets would later come under fire from feminists and sex educators for their obfuscation of the connection between sex and reproduction, at the time they were first produced it was exactly this separation of menstrual education from sex education that made them so valuable. Parents were much more willing to educate their daughters early about menstruation if they could avoid talking about sex, so for many girls, the availability of these pamphlets made the difference between having some explanation of menstruation and receiving no information at all.

Kimberly-Clark's first pamphlet, *Now Science Helps Women Solve an Age-Old Problem*, was signed by nurse Ellen Buckland, who also appeared as an expert in Kotex advertisements and was responsible for correspondence about the pamphlets through at least the mid-1930s. This upbeat 1923 pamphlet, aimed at adult women, did not give much explanation of menstruation but rather promoted Kotex as "the New Hygiene which is bringing comfort, peace of mind and greater efficiency to the world of women." It explained that using absorbent, disposable Kotex prevented the dangerous growth of bacteria that it claimed was likely to take place on cloth napkins. It also touted Kotex as scientific in its engineering, "far superior to the old-time, makeshift ways." It encouraged women to regard menstruation as "a natural function" and continue with most activities during menstruation, since "periodic disability is out of keeping with modern advancement." Regarding menstruation as normal, continuing regular activities, and using Kotex to make this possible was supposed to create "happy, well-poised, efficient modern women."[25]

Buckland's advice celebrated the values of "new women," seeking advancement and personal fulfillment through engagement in the public sphere. At the time she was writing, in the early 1920s, the reformers and working girls of the Progressive era were being overtaken by a novel variety of new womanhood: the "flapper." Both white and black and mostly middle class, flappers appropriated the innovations of working girls of the previous twenty years, embracing sexualized self-presentation and mixed-sex entertainment. They went to college, like the new women reformers who came before them, but they were much more likely to marry, parlaying their flirting, drinking, and jazz dancing with men into companionate marriages with men they expected to serve as friends, confidants, and romantic partners as much as economic supporters. While less visible than the reformers in the public world of political and social reform, they were highly visible in their leisure-time escapades out on the town. Flappers set the tone for an American culture newly enamored of youth and glamour, and they became role models for working girls who aspired to join the middle class as well as younger girls aspiring to become alluring college coeds.[26] Buckland captured the buoyant spirit of the "new women" of the early twentieth century and tried to promote Kimberly-Clark as a key supporter of new women, whether reformers, working girls, flappers, or aspirants. Kimberly-Clark, as represented by Buckland, was a company willing to be frank about a topic widely regarded as sexual and to give women the technology and moral support necessary to happy new womanhood.

In 1928, Kimberly-Clark hired Dr. Geoffrey Williamson to write another pamphlet for adult women. This one gave a more scientific explanation of menstruation, explaining that it is "the natural discharge of a bloody fluid from the woman and occurs about once each month . . . carrying small bits of lining that are thrown off from the inside of the womb at this time." While he encouraged women to regard menstruation as "a perfectly normal, natural function of the healthy woman," unlike Buckland he detailed a range of menstrual complaints, concluding that "entire absence of suffering at the menstrual time is known only in sixteen women out of a hundred." Although he gave ideas for relieving these aches and pains, he seemed to have a much more pessimistic view of the possibilities for "modern woman."[27]

Around the same time, a Kotex pamphlet published in Australia, aimed at girls of eleven or twelve years old, emphasized the need to take it easy during menstruation to avoid nervous strain, injury, or catching a cold as a result of lowered vitality. While this health advice was even more out of keeping with Buckland's image of the "modern woman," the pamphlet broke new ground in targeting prepubescent girls and explaining to them that menstruation was "a flow of blood, lasting for several days . . . discharged through the vaginal opening of the body . . . caused by a breaking down of blood cells in the inner lining of the uterus."[28]

This Australian pamphlet must have struck some readers, or at least some Kimberly-Clark executives, as far too revealing for the preteen reader, sexually frank to a degree that might have been fine for "new women" but not necessarily for their young daughters. Kimberly-Clark's next pamphlet, *Marjorie May's Twelfth Birthday*, first published in Great Britain in 1929 and in the United States in 1932, took a much more conservative approach. This booklet narrated the story of a mother telling her twelve-year-old daughter about menstruation. The mother explains that menarche initiates "a new method of purification," comparable to defecation. She then shows her daughter that Kotex can absorb this "wonderful purification which performs to keep your new physical development free from waste."[29] This explanation of menstruation probably struck many mothers as commonsensical and pleasant, giving a positive spin to the traditional explanation of menstruation as a way to rid the body of "bad blood," and also as an easy way to talk about it with a young girl. However, it was a striking turn away from Kimberly-Clark's previous attempts to be up to date and scientific. This pamphlet, with its surprisingly traditional explanation of menstruation, was the pamphlet that Kimberly-Clark advertised and distributed widely for pre-menarcheal girls for almost a decade.

At the same time, in 1933 Kimberly-Clark published yet another pamphlet for adults, by Dr. Lloyd Arnold, which almost sounded like it was scolding Marjorie May's mother for her traditional explanation of menstruation. "There is a mistaken notion that this is a cleansing process. It is *not* a cleansing process. There are no more impurities in a woman's system than in a man's." Rather, "There are periods of time in all female animals when the womb or uterus is especially prepared for receiving the cell to be fertilized. The monthly menstruation is part of this period in a woman."[30] In writing his pamphlet, Arnold took advantage of the scientific consensus that had just been consolidated about the timing and purpose of menstruation in relation to ovulation. By the early decades of the twentieth century, pathologists had become better able to tell the age of a dissected corpus luteum, making the timing of the ovulatory and menstrual cycle more possible to discern. The new ability to visualize and locate an unfertilized egg, beginning in 1930, allowed scientists to pinpoint this relationship more accurately, supporting a consensus that ovulation occurred about fourteen days before menstruation. If this was the timing of ovulation in relation to menstruation, clearly menstrual bleeding was not the equivalent of estrus bleeding in dogs.

Additional research into the release of hormones by the corpus luteum also made it clear that ovulation triggered the buildup of the uterine lining, and that if the egg was then not fertilized, the consequent hormonal withdrawal set into motion the sloughing of the uterine lining and the bleeding of menstruation.[31]

Arnold spent two pages describing in clear detail the anatomy and physiology of ovulation and menstruation, concluding, "So, you see, menstruation is merely the throwing off of material that has gathered in the womb but has not been used. There is nothing unclean or impure in this natural process."[32] Like Samantha, who read this sort of explanation in her school library and found it enlightening and reassuring, Arnold believed that a scientific explanation was the best remedy for negative feelings about menstruation based in a traditional understanding of its purpose. This lengthy scientific explanation served as the springboard for the health advice in the pamphlet, in which Arnold urged women to see even a multitude of minor discomforts of menstruation as normal and in no need of special treatment or rest.

A few years later, Kimberly-Clark issued a new pamphlet aimed at teenagers in which *Marjorie May Learns about Life,* or more precisely, learns how babies are made. In this pamphlet, author Mary Pauline Callender ignored her previous explanation of menstruation as purification and instead explained that "when one is not married and there is no baby growing in the womb, nature needs neither the ovum nor the menstrual fluid, and they leave your body through the same passage a new baby would use in coming into the world."[33] Taken together, the two Marjorie May pamphlets and Dr. Arnold's pamphlet for adult women demonstrate the dilemma that Kimberly-Clark and many mothers shared: they thought girls should know about menstruation before menarche, and they valued a scientific explanation, but they really wanted to avoid talking about sex until high school age. Kimberly-Clark's solution for many years was to provide at least some sort of explanation to twelve-year-old girls, even if it was not scientifically accurate, and then replace the traditional explanation with the scientific one later, in the context of sex education.

Through the 1950s, this pattern of producing separate booklets for pre-teens and teens persisted, in the literature produced both by Kimberly-Clark and by Johnson & Johnson, which made Kotex's primary competitor, Modess sanitary napkins. Johnson & Johnson published two pamphlets in the late 1930s, one for pre-menarcheal girls, *What a trained nurse wrote to her young sister* and one for teens, *The Periodic Cycle*. The forward in *The Periodic Cycle* made it clear that this tiered approach had been carefully planned. It explained that all girls should be told about menstruation before menarche. "But most girls, when this first information is given to them, are too young to understand fully all that the menstrual function means. Later in their teens they should be given more detailed instruction, under a teacher's guidance." "What a trained nurse" told girls to expect was "something new—a slight flow of blood from your body. And you should be very happy, because it means that your body is developing as it should, and that you are well on your way to being a real grown-up!" Following the example of *Marjorie May's Twelfth Birthday*, in its reticence, this pamphlet failed to even mention from where the blood would flow, though presumably a girl could infer the general location from the pamphlet's instructions about how to use Modess on its final pages. And it avoided giving a scientific explanation by downplaying the recent advances in understanding of the cycle. "Even the wisest doctors still argue about precisely what starts menstruation and what is its purpose. But they're all agreed on one point—it's a thoroughly natural and normal thing and nothing to worry about for a minute."[34]

In contrast, *The Periodic Cycle* gave an extensive description of the menstrual cycle, describing ovulation, the development of the Graafian follicle, and changes in the uterus leading to menstruation. Unlike *Marjorie May Learns about Life*, it did not attempt even a vague description of the sex act but stated that menstruation happens if there is no "fertilization (union of the male and female cells)."[35] These two pamphlets together illustrate clearly the plan for menstrual education which manufacturers and parents seemed to find most comfortable through the first half of the century: Warn pre-menarcheal girls that it will happen, without explaining much, and then give teens a scientific explanation of the reproductive cycle with as little discussion of sex as possible. While less than ideal from a sex educator's viewpoint, this approach of separating menstrual from sex education allowed manufacturers to produce broadly acceptable, uncontroversial educational materials that they felt enhanced their authority and their customers' brand loyalty, and gave parents explanations they were actually willing to share with their daughters.

Just after World War II, inspired by the outpouring of short educational films created for soldiers and civilians and by their own extensive participation in the Women Army Corp's program to reduce work absences due to menstrual difficulties, Kimberly-Clark produced a ten-minute film about menstruation for school audiences.[36] Created by Disney, *The Story of Menstruation* gave cartoon illustrations of the menstrual cycle as well as health and hygiene advice. The information was framed by a narrative of a pubescent girl's happy daydreams of going on dates, having a wedding, and caring for a baby, the life trajectory that menarche was supposed to set in motion. A new pamphlet, *Very Personally Yours*,

54 The Modern Period

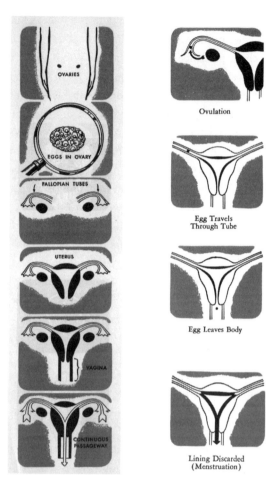

Stills from the 1946 Disney educational film *The Story of Menstruation*, reprinted in Kimberly-Clark's pamphlet *Very Personally Yours*

was created to accompany the film, and basically reproduced its information and illustrations in written form. Initial advertising for the film did not specify for what age group it was meant, though it quoted happy teachers who had shown it to seventh and eighth graders.[37] The teaching guide that could be requested to accompany it encouraged teachers to address menstruation informally with fourth graders and then show the film in seventh grade.[38] According to a 1964 memo from Kimberly-Clark's educational department, teachers quickly began showing the film to younger students in the fifth and sixth grade.

In 1952, Kimberly-Clark added another pamphlet for these younger girls, *You're a Young Lady Now*, which gave a much briefer but still scientifically accurate description of the menstrual cycle and its relationship to reproduction. While younger girls were supposed to receive the pamphlet with the watered-down explanation of menstruation, they still saw the film, and the line between information appropriate for twelve-year-olds and for teens was blurred. Finally, by 1957 Kimberly-Clark was urging schools to show *The Story of Menstruation* to fifth graders, sending teachers and principals reprints of an article about how a Catholic school began showing the film to its fifth- and sixth-grade girls, gaining the full approval and appreciation of the girls and their parents.[39]

The Story of Menstruation fit comfortably within a 1950s culture that emphasized sexual "containment," the promotion of sexual satisfaction for men and women alike within marriage and in service of the stability of the nuclear family and the suburban household.[40] "Family life" curricula proliferated, teaching girls and boys how to prepare for and achieve this cold war update of the companion-

Kimberly-Clark publicity photo demonstrating the use of *The Story of Menstruation* in a Catholic elementary school, reprinted from *The Catholic School Journal*, January 1957

ate marriages of the 1920s.[41] While in fact, African American mothers had never left the workplace and white mothers were increasingly entering it, the wife as homemaker was glorified and idealized.[42] Birth rates temporarily increased, causing a baby boom. Unlike their nineteenth-century predecessors, even mothers who could afford to stay home with their children in the 1950s often had their three or four children in rapid succession in their twenties and then ceased childbearing altogether, opening up space for a full-time career outside the home after the children were in school or had departed for work or college. For many families, the income women earned before they had children and after their children entered school was crucial in establishing a suburban, middle-class lifestyle. Homemaking, marital sexuality, and childbearing were ideologically central to postwar America, while mothers' income from work was necessary for many families to achieve this middle-class domestic ideal.[43] Kimberly-Clark's educational efforts, so supportive of "new women" in the 1920s, by the postwar period had been reframed to comfortably align with those cultural commitments. Its film and pamphlets linked menarche to a dreamy domesticity while maintaining the practical advice girls and women sought as they navigated schools and workplaces, as well as dates and honeymoons.

Kimberly-Clark's education department distributed millions upon millions of pamphlets over the years, through schools, workplaces, and also by direct mail. A 1964 department memo indicated that *The Story of Menstruation* alone had been seen by 47 million girls by that time, and *Very Personally Yours* and *You're a Young Lady Now* had together been distributed to about 31 million girls. In that year, Marion Jones, director of the education department, estimated that Kimberly-Clark's materials reached 20 percent of American girls, and she requested more funds to try to reach the rest.[44] By the time it was replaced by the film *Julie's Story* in 1984, *The Story of Menstruation* had reached approximately 100 million viewers.[45] The distribution numbers for earlier pamphlets are less well documented but reached into at least the several millions for *Marjorie May's Twelfth Birthday* and *As One Girl to Another*, a 1940 update written by *Ladies' Home Journal* sub-debutant (pre-teen and early teen) editor Elizabeth Woodward in the chatty, casual style of a teen magazine.[46]

Kimberly-Clark's pamphlets reached so many girls because they were distributed in a multitude of ways. Unlike other sex education materials, which were primarily available through schools and in some libraries, these pamphlets were advertised widely and could be obtained for free by mailing in the coupons included at the bottom of many Kotex advertisements. Other manufacturers followed Kimberly-Clark's lead, and a young woman or her mother browsing through a women's or teen magazine anytime between the 1930s and the 1960s would have had a good chance of finding an opportunity to order a pamphlet explaining menstruation. *Marjorie May's Twelfth Birthday* was even sometimes included in Kotex boxes, or as a free handout from retailers.[47] Coupons to send away for free pamphlets were also included as part of the instruction sheet in Kotex boxes, even when the pamphlet itself was not included.[48] With constant reminders of their availability and a cheap, easy, and private way to obtain them, manufacturers' menstrual education pamphlets were revolutionary in extending the reach of rudimentary sex education for girls. In the early 1960s, studies conducted on college freshmen by sex educator and researcher Warren Russell Johnson showed that young women were clearly better educated about sexual anatomy and physiology than young men. Johnson attributed this difference to menstrual product manufacturers' educational programs, concluding that "their advertising has encouraged a popular acceptance of at least a little sex education for girls."[49]

Some historians and feminists have criticized the role of commercially produced pamphlets in menstrual and sex education, dismissing Johnson's perspective. For example, historian Joan Brumberg argues that manufacturers such as Kimberly-Clark took over the mother's job of educating a daughter at menarche and in doing so, changed the meaning of menarche from an entrance into womanhood and adult sexuality into a hygienic crisis solvable through the purchase of the right hygiene products.[50] Brumberg's vision of the ideal menstrual education, in which a mother or other trusted advisor teaches a girl about her body, her developing sexuality, and its relation to her role as a woman certainly has a great deal of merit, especially if that older woman acknowledges a wider range of possibilities than the single life trajectory represented in *The Story of Menstruation*. Historically, however, there is no evidence that mothers in America ever routinely educated their daughters in that way. Like Ida Smithson, older interviewees who embraced manufacturers' pamphlets as a learning tool were comparing them to the available alternative: terse and incomplete advice on hygiene and health from their parents—provided in response to girls' initial shock and fright at finding themselves bleeding—or whispers from their classmates.

Ida and others expressed their belief that the communication initiated by providing the booklet was highly valuable, a real step forward in mother-daughter communication. As Ida expressed it, "once you keep confidence in them, it's far different than if they've got to learn it from somebody else." Ida gave pamphlets to every one of her daughters, from the 1950s through the 1970s, likely including *Very Personally Yours*. Rachel Cohen took a slightly different approach; before she reached menarche in Denver in the early 1940s, she ordered her own pamphlet.

It was in some magazine or someplace, Tampax used to have an ad, and the ad said if you send ten cents, you get a little booklet. You had a little coupon you filled out with your name and address, and for ten cents, you get a booklet entitled, *What Every Girl Should Know* . . . I was very uncomfortable knowing that this was kind of hanging in the air, and I just wanted my mother to know that she doesn't have to kind of pretend it doesn't exist. I just didn't feel comfortable about it. It was more just to sort of make a formal announcement and have something to make it about than for my own knowledge, because I knew what it was all about . . . And then when I got it, I said, "Oh, Mom, I got this," and I announced it to her, and she just kind of brushed it off.

Despite that Rachel had gathered much of her information about menstruation from openly overhearing conversations between her mother and her mother's friends, she still felt that her mother did not consider menstruation to be an open topic. The booklet did not exactly open up a wide avenue of communication between Rachel and her mother, but it was a way for Rachel to insist that menstruation could actually be discussed in ways other than whispers and complaints among women friends.

Many older interviewees found the communication they achieved with their mothers and daughters using pamphlets quite satisfactory. Only as this level of communication began to be taken for granted did interviewees express guilt as mothers or disappointment as daughters if menstrual education consisted solely of a pamphlet and some reassurance. Some, like Samantha, were able to immediately adopt sex educators' model of openness about menstruation and sexuality, but more typically interviewees born in the 1930s and 1940s found that they wanted to be more open with their daughters than they actually managed to be. Liza O'Malley had received a pamphlet from her mother and in turn gave pamphlets to her daughters. She felt well prepared for menarche herself, having learned from the pamphlet and her friends at school, but she felt guilty about having just handed her kids pamphlets, in the 1960s and 1970s, without an extended discussion about sex and reproduction.

I was always waiting for the right time. And I think they were waiting for the right time. Or maybe they were waiting for me. And I just wasn't able to. They would joke about it, you know? Recently. The oldest one said, "Yeah, we're still waiting for our sex talk with Mom." And I said—my mother was alive at the time—I said, "Yeah, when I have mine with my mother, then I'll sit down and have one with you." [Laughter.] . . . I wished

I had a more open relationship, but I just couldn't do it. I was always waiting for the right time and it never happened . . . Many of the things that I said I didn't want to be like my mother, I was.

While Liza felt that she should have provided even more information to her daughters, she and her mother at least provided a basic education about menstruation to their daughters before menarche, with the help of a manufacturer's pamphlet. By the late 1950s, when Kimberly-Clark was urging even Catholic schools to show *The Story of Menstruation* in the fifth grade, pre-menarcheal menstrual education seemed much more likely than not to happen. All interviewees born in America after World War II were unsurprised when they got their first periods. Still, this pattern was not yet universal. Those who did not have even the modest resources with which so many Americans were becoming "modern" were perhaps the most likely to be left out of this new pattern of menstrual education. African American Laurie Wilson was born in the rural South to a very young mother in the early 1950s and was raised primarily by her grandmother. She recalled her experience in the mid-1960s: "I remember being so glad my grandmother had prepared me for that, because some of my friends, this did happen to them at school, where a teacher would have to carry them home—at that time, there were not a lot of telephones in the households . . . A lot of them, you could tell they were terrified, because they didn't know what was happening."

American parents, along with primary school teachers, sex educators, and pamphlet writers, eventually made pre-menarcheal menstrual education almost universal in the United States, and they created the scientific language and pedagogical circumstances for talking about menstruation without shame or embarrassment. However, it was not a straightforward path, despite widespread agreement that this education was appropriate and desirable. To get there took more than a half-century of sincere but halting, awkward efforts in that direction, as individual women struggled to instantiate the new ideals they set for themselves, and social institutions such as schools and menstrual product companies revised curricula and created a public discourse upon which women could draw.

Among women interviewed for this book, those born in the 1940s and 1950s seized upon the scientific explanation of menstruation even more avidly than their predecessors and were more active in seeking out further explanation after the first mother's talk, manufacturer's pamphlet, or grammar school education program. For example, Laurie Wilson explained that "when it first started at twelve years old, it was no big deal. But then, when you get older, you want to know the reason your body is going through all these things, and what is really happening." She read books from the library and articles in teen magazines, and paid attention in high school health class to get "a lot more in depth information about it. Maybe things your grandmother was not aware of . . . Because she could only relay to me what she knew about it. Since she had not been able to go to school, it was just whatever *her* mom had told *her*."

Women born in the 1940s and 1950s were also likely to learn more about menstruation in college. In the 1950s, courses and majors aimed at teaching women a particular middle-class feminine role, based in homemaking and family caretaking, became widespread and popular, expanding from a base in "progressive" education for women instituted in the 1930s.[51] Learning the scientific approach to understanding human sexuality and reproduction was considered important, both to support a "rational" approach to marital sexuality and family planning, and to enable women to feel comfortable educating their children about sexuality and reproduction.

New Englander Barbara Ricci remembered how much taking these classes in the mid-1960s changed her perspective. She talked about how she came to use the term *menstruation*, even though she had heard and used other terms growing up.

> I learned to be less self-conscious in college than I did growing up talking about the male and female parts of the body, like saying the word *penis*, and *erection*, and things like that. Because we used them in a very, almost, like, technical way, it then made it easy to talk about lots of bodily functions and parts of the body. And since I was studying them, and maybe even learning how to teach other people or tell people about them, it just made it easy to use all the words. And there was nothing embarrassing or secretive about it; they're all normal, human functions and parts, that everyone should be able to talk about . . . I think it was a modern way of looking at it and talking about it, that would have probably shocked the general working class, that we were learning about this, that there was a *class* on it. They probably would have joked about it. So I think it was a very modern way of learning about it. And very valuable for me . . . College is a way of opening your whole horizon. For me that was the most valuable thing about going to college, was to see all of these ideas and possibilities, that were so different from how I grew up.

Barbara valued the way her classes made menstruation into an acceptable topic for discussion, and a topic worth discussing at length, although she knew her high-school-educated parents, who worked as waiters, and the other members of her childhood community, would have found it ludicrous. For women who were the first in their families to adopt "modern" norms surrounding menstruation, college was often the place where they most clearly made and recognized a break with traditional ways of thinking about and managing menstruation, which they often saw as tied to their parents' less-educated outlook. Barbara explicitly related her self-conscious entry into a middle-class world view to her adoption of "modern" ways of thinking and talking about her body. Among those interviewed for this book, women from a tremendous range of backgrounds, including Barbara and Laurie, adopted a modern perspective that would facilitate their entry into an extremely broad self-perceived American middle class.

When Barbara, Laurie, and other women of their generation had their own children in the 1960s, 1970s and 1980s, they were by and large ready and willing to educate them about menstruation. However, when it came to giving a scientific explanation of menstruation, they often found themselves preempted, or at least upstaged, by school programs. By the 1990s, at least 88 percent of elementary and junior high schools offered education about menstruation.[52] School programs also became more likely to include boys, though through the end of the century it was still common to segregate education about puberty and menstruation by gender. Groups such as Planned Parenthood developed materials and had on-staff teachers who went to the schools specifically to teach sex education, developing and professionalizing these school programs. The teaching methods this generation remembered were more varied and elaborate than their mothers recalled, drawing on the much greater range of curricula available. Among interviewees for this book, only those who went to Catholic schools did not get any school education until at least eighth grade, which was probably after menarche for about half the girls.

Pamphlets continued to be an important source of information as well. Jennifer Kwan, who emigrated with her family from Hong Kong as a child, distinctly remembered the one she read.

> Even before I saw the film, I had a friend who got something from her pediatrician, something about "Menstruation," and it was this kind of

> dorky pamphlet that was from one of the pad makers—I think it was Kotex or something. It had all these flowers, and birds . . . stuff like that, and so I had a little handout about that and I was fascinated by it . . . That offered some explanation, a little bit better than what we got in the film, even.

Felicity Chang, the daughter of immigrant restaurant-owners, "kept like a stack of them in my drawer, and this is going to be pretty funny . . . I kept them as reference, because I knew that at one time I would need them, and I wasn't that keen on reading all of them at that point."

Among interviewees, those growing up in the 1970s, 1980s, and 1990s were able to take for granted innovations in areas much broader than just early menstrual education. They took for granted the rights achieved by the feminist and civil rights movements, including opportunities to play sports as determined by Title IX, the increased control over their bodies afforded by access to legal abortions, rising female enrollment in colleges and professional schools, and legal (though not always practical) protection from sex- and race-based discrimination in the workplace. They assumed that they would have both career and family, and that they would be given the intellectual and practical tools with which to achieve their goals. Early menstrual education, which gave them scientific and practical information to help them manage their bodies, was one small step along the way.[53]

Between the menstrual education programs in fifth-grade classrooms and the manufacturers' pamphlets, mothers who tried to tell their daughters about menstruation in the last decades of the century often found that the job had already been done for them and their daughters considered the information old news. Joan Randolph, whose father was a professor and whose mother owned a series of small service businesses in New England, related what happened after she got her first period in the early 1980s.

> I remember my mother saying to me, "So"—such a mother kind of thing to do—"So, do you have any questions about what happens when you get your period? You know, do you know what happens?" I said, "Oh yes, it's the sloughing off of excess material after the egg didn't get fertilized." My mom's like, "Ah, OK! Obviously we don't need to have a discussion about this."

Mimi Cummings Wood, younger sister of Roberta Cummings Brown, talked a little with her daughter, Lisa, but "not in great length, because you know, I found out that a lot of times, some of the subjects, your children really don't want to get that involved. 'Oh I know already, Mom.' So what are you going to say? 'Oh, well I insist!' [Laughter.] 'I've been knowing that. I've been told that. I've been knowing that forever.'" Lisa was not very interested in hearing her mother's educational spiel about menstruation, an education that Mimi had felt terribly deprived of herself and wanted to share with her daughter. Still, Mimi's attempts at educating Lisa about menstruation opened up room for later conversations in which they shared their experiences with each other. Lisa complained to Mimi when she suffered from cramps, and Mimi told Lisa stories about how awful her cramps had been as a teenager, reassuring her that they would subside as she got older.

While many daughters who reached menarche in the last decades of the century may have lightly dismissed their mothers' attempts at menstrual education, some of those whose mothers did not make the attempt felt very disappointed by the omission. New Englander Carlen Joyce Thomas expressed the daughterly disappointment that seemed to be the counterpart to Liza O'Malley's motherly guilt. She explicitly regarded Judy Blume's coming-of-age novel *Are You There God? It's Me, Margaret* and the booklet with "a pink cover" she received, which she "read cover to cover," as a substitute for her mother's guidance. "My Mom did not discuss much of it, so for me, that was my source of information . . . I didn't feel comfortable going with questions to my mom. So I think whatever I could get my hands on bookwise, in the library, I think that's what I relied on." Carlen's mother, Dorothy, a nurse, felt that she had made herself available, accompanying Carlen when she viewed the educational film about menstruation and preparing to answer questions, but since Carlen waited for her mother to open the conversation, it never took place. Carlen felt that her mother had passed up an important opportunity to initiate open conversations about sexuality, something she hoped to do differently with her own children. She believed that a discussion of menstruation between mother and daughter "sets the stage for talking about lots of other things, and talking about your feelings. Talking about sex, and talking about relationships." She regretted not having felt comfortable talking about sex and relationships with her parents and thought that an extended discussion of menstruation would have created an opening for talking about these things.

Not everyone shared Carlen's expectations, though. Young Chinese American interviewees born in the later decades of the century seemed to have particularly low expectations about how much they would discuss any of these matters with their mothers, so they were more satisfied with the pamphlets and school programs as a substitute. Amanda Chen, daughter of an aerospace engineer and computer scientist, mused, "I think she assumed we learned it at school, because they do sign little permission slips that say you can watch the video. I think she assumed that we'd learn it at school and it would get taken care of. Same with sex ed in general." Heidi Xue, growing up in a successful entrepreneurial family in the suburbs of Los Angeles in the early 1980s, was satisfied with this approach. "You know, in junior high, you're made to watch those films, when you're, I think, 11, or something like that, to prepare you. I guess we both assumed that everything was said in that, so . . . I didn't really think I needed any more information than that." Jennifer Kwan, who was a full-time sex and menstrual education teacher, would have preferred a fuller explanation but sympathized with her mother's position. "I got a little briefing about it, but otherwise, not a whole lot from my mom. Of course, my grandma never talked to her about it."

Like Jennifer's mother, interviewees who had grown up in midcentury China were likely to have been surprised themselves at menarche and received little instruction or support when they did get their periods. Families were on average larger than American families, and while sisters did not necessarily talk much to each other about menstruation, a sister was likely to help handle a first period and explain how to manage the flow. Laura Hwang, the daughter of an army general and a homemaker, explained why her twin sister came to her when she got her period, even though Laura had not gotten hers yet and did not have any knowledge about managing it. "Chinese people, like my mom's generation, think that's a dirty thing; she doesn't want to talk about that. So we tried to figure out how to do it. They think we automatically know how to take care of it." After that, menstruation was not mentioned unless a girl had menstrual pain. As Phoebe Yu, the daughter of a government official and a homemaker, put it, "It's just my sister taught me how to do it, and I just did it. And from that point I assumed that's the way it is." Florence Wu, from a well-off Hong Kong family, had a similar experience at menarche, and was surprised and touched by an experience as a volunteer at the nurse's office at her daughter's elementary school in California. A girl had come in complaining of stomach cramps.

> "I said, "Why don't you go and check [if you have your period]. I'll give you this pad." She came out, and she said, "Oh yes, I got my period! I've got to call my mother." And that's the first thing she did! She called her mother from the office, and she told her mother that she had her period! And I thought, "How nice, to share this with your mother." And I remember the first time I had it I was so traumatized by it all . . . and I could not even go to my mother and tell her."

Almost all of the women interviewed who grew up in China learned something about menstruation several years after menarche, in high school biology.

Many remembered, however, that when they reached the chapter on human reproduction, as Betty Li put it, "the teacher would say, 'OK, you go home to read it.' [Laughter.] I don't remember that I went home to read it, either! Come in and sort of forget about the whole thing!" Not everyone forgot about reading the chapter; Laura Hwang remembered it as a key source—in fact, her only source: "Just that schoolbook, about your health, about your body. And the last chapter about man and woman, about the symptom, and about the breasts, and the period, and the hair, something like that." She laughed about her sister, who must have failed to read that crucial chapter.

> My twin sister came to United States when she was twenty-three, I believe. At that time she had a boyfriend, that right now is her husband. Boyfriend try to kiss her, and she was so panicked, pushed him away. He say, "What happened?" "Oh, I don't want baby!" [Laughter.] She think that kissing will have baby, because all the early day movie . . . "Oh, when they have kiss, somebody will have baby!" [Lots of laughter.] Twenty-three years old already, and she doesn't know that!

Laura and the other immigrant Chinese mothers interviewed almost uniformly wanted their daughters to have more information, earlier, than they had received themselves. They found the scientific explanation they read in a high school biology text or picked up later in the United States appealing and useful compared to the terse instructions about menstrual hygiene they had gotten at menarche from sisters, household servants, or mothers. They were not necessarily confident about imparting it themselves, but they were enthusiastic supporters of this aspect of the "modern" American approach to menstruation.

Phoebe Yu found that she was able to provide this kind of "modern" explanation on the spur of the moment, despite not having prepared at all, given that she only had sons. She remembered a situation in which she was forced to say something to her young sons, in the mid-1990s.

> There was one time, I think I made a mess on my nightgown, and I didn't know, and they noticed, and they were asking why, what is it. Did I get hurt or anything. That was kind of hard for me to explain. I explained to them, I said women do have the cycle, to prepare their body for having a baby, and so—what did I tell them? If you end up not having the egg, if those eggs were not used, then you have to get rid of them, and that's why. You get rid from the body, you get rid of them, that's what the blood is coming from.

Phoebe's explanation to her sons was remarkably different than the curt instructions she was given herself at menarche. She, like so many Americans of her generation, had become fully invested in the "modern" way to teach about menstruation.

―⁂―

Whichever combination of sources they drew upon to learn about menstruation, the youngest women interviewed from all backgrounds were quite confident in their scientific knowledge of the menstrual cycle, which most of them had learned long before menarche. Able to take for granted the availability of this information and the positive affect with which it was imbued, they began to critique it. They found fault both with what it included and what it left out. They certainly were not rejecting the modern approach their mothers had embraced, but they demanded more from it. They wanted to refine it to provide even greater ability to manage their bodies effectively and without anxiety.

Traci Anderson, daughter of a white New England professor and a special needs teacher, wondered if the scientific story really addressed a pre-menarcheal girl's concerns. She described how she would explain menstruation to a ten-year-old girl.

> I guess more than explaining the technicalities, I would try to just be comforting and explain that it's different for different people, and that it changes quite a lot during the course of your life, and that you should be able to tell people if it hurts and get the sort of help you can get for it. More try to be reassuring than anything else, and tell her it was scary when I started it, and it's always scary when it starts. Maybe not go into the details of the science so much because that's probably not what they're worried about.

Amanda Chen liked the scientific explanation but thought it was inadequate. When she got her first period,

> I wasn't sure that that was it, at first, because it wasn't what I expected. It was kind of like, brown, and I thought, "That's weird." They always describe it as blood, so I thought it was like blood when you cut your hand, which is a very different color and texture and everything about it is just different than menstrual blood. And I thought of course it would be the same. So I wasn't really sure what was going on.

Joan Randolph found her school's very thorough menstrual and sex education lacking in another way.

> One of the funny overeducated, but undereducated, moments I remember is my mother getting out of the bath, and I was in the bathroom one time, and obviously what it was is that she had a tampon in, because I could see the string, but . . . I knew more about IUDs than I did about tampons, and I decided it was an IUD. I didn't know jack about how it worked, but I knew there was a string. No one ever had ever taught me anything about tampons.

Traci, Amanda, and Joan all pointed to what they felt were serious lacuna in their menstrual education, flaws that could be seen clearly against a background of taken-for-granted basic, early education about the science of the menstrual cycle.

Another important critique implicit in many young interviewees' comments was the reproductive focus of the standard scientific explanation of menstruation. If the point of the menstrual cycle is reproduction, then what use is the cycle to a preteen? According to Felicity Chang, "I thought it was a huge inconvenience, and I really didn't see the point of it. I always wanted to have kids, but I wouldn't want to have kids for a while, and it annoyed me that I had to deal with it." While their mothers and grandmothers found the scientific explanation, focused on reproduction, a positive and reassuring alternative to traditional ways of explaining menstruation, this younger generation of women clearly felt the limitations of this approach. As women who were unlikely to be trying to have children until at least a decade, perhaps two decades, after menarche, they were not excited to have hundreds of what they thought of as "useless" cycles before they got pregnant.[54] One 1975 Kimberly-Clark pamphlet for older teens, *Getting Married*, reworked the scientific narrative in a way that might have given these young women a more positive feeling about menstruation. It began with the standard narrative that menstruation occurred if a fertilized ovum did not arrive in the uterus. Then, it concluded, "the lining of the uterus once again is pink and smooth and nature has completed a miraculous process of renewal."[55] This revised narrative only appeared once, however, and never became part of the standard explanation.

While women born in the later decades of the century critiqued the scientific explanation of menstruation, they were able to find many more alternate published sources of information about menstruation than had the generations before them. Judy Blume's book, *Are You There, God? It's Me, Margaret*, first published in 1970, was very important for girls and also for boys who had sisters and borrowed the book from them. In Blume's book, four fifth-grade girls form a club to talk about boys, breast development, and periods, which none of them have gotten yet. The girls wonder if others will develop before they do and want to know what getting a period really "feels like." They experiment with buying sanitary napkins, and the narrator tries them secretly before she gets her period.

Blume addressed the anxieties of pre-menarcheal girls, with a story that assured the reader that she was not alone in feeling anxious or nervous about menstruation, and simultaneously offered the kind of reassuring practical information about menstruation that often supplemented the scientific story in manufacturers' pamphlets and school programs. For example, a reader could learn from Blume's book that periods could be irregular at first, that there is a wide range of ages at which girls can get their first periods, and that menstrual pads could be worn with a belt (1970 edition) or attached to underwear with a sticky strip (1991 edition). Blume also poked fun at school education programs, at the same time that she assumed her readers had been exposed to them, the narrator referring to a film that "told us about the ovaries and explained why girls menstroo-ate."[56] Blume's characters, like the young women interviewed for this book, knew a lot about menstruation before menarche; their anxiety was focused on wondering when it would come, and what it would feel like. One of Blume's characters describes to her friends what it felt like to her to get her first period and how she handled it, giving readers more experiential information about menstruation and a model of sharing this kind of information with friends.

Young women valued this perspective on menstruation that was mostly missing from formal education programs. Rose Mitchell, daughter of an engineer/manager and homemaker, recalled that she and her friends "were all reading *Are You There God? It's Me, Margaret*–type books. So I was constantly thinking, 'Oh, I think it's happening now,' and rushing to the bathroom." She was not alone; almost all the young women interviewees had read the book and mentioned it as a primary source of early information. Heidi Xue pointed out that she got information from "the Judy Blume books we grew up reading, but no, I never felt the need to announce it to anybody. [Laughter.]"

It was not only girls who combined *Are You There, God? It's Me, Margaret* with educational pamphlets to learn about menstruation. Chinese American Tim Dai, son of an engineer and an elementary school teacher, could not compare his own experiences to Judy Blume's story, but he found it a useful source of information anyway.

I think one thing which was probably one of the earliest exposures I had to menstruation was, again, by virtue of having a sister, there were things around the house. So I think my sister was reading certain kinds of books at a certain age, and I think I just ended up reading them, too. So I'm not sure they're really written for people like me, but I think I read them anyway. I think those were like Judy Blume books and things. I think that that was probably some of the first exposures I really had.

While these young women and men were growing up, television and radio got into the act as well, providing practical information, jokes, and ideas about menstruation. Beginning in the 1970s, in keeping with the greater openness about bodily functions promoted as part of the sexual revolution, ads for pads and tampons were shown on television. Appearing rather belatedly compared to talk about sex, jokes and casual talk about menstruation—which previously had been private and largely shared among single-sex groups—were put into sitcoms and pop songs in the 1980s and 1990s. Young men were particularly conscious of how much they learned about menstruation from the mass media. Tim Dai explained, "I feel like I have gotten a lot of exposure to a lot of products associated with menstruation through television. And my personal feeling is that I feel like I have a detailed knowledge of how these things are designed from watching TV commercials and I'm a little bit struck by that, because TV is everywhere." Like Tim Dai, Keith Zhao, son of a physician and an accountant, felt that "I get a lot of exposure to, kind of, how it's treated, culturally, just by watching television, watching movies and stuff like that." Alex Jones, the son of African American educators, remembered movies, and hip-hop and rap albums, which discussed menstruation much differently than his school sex education program, offering "another definition of what was taking place." Movies such as *New Jack City,* with "city themes," Alex recalled, would show a couple fooling around, "and the girl would be like, 'I can't right now,' or something." His friends commented on it, saying, "Oh, she's in the red zone," or she had a "girl problem." In 1988, the sitcom "Married . . . With Children" produced an entire episode full of period jokes, only fifteen years after television censors had reluctantly begun to allow very restricted airing of commercials for menstrual products.

While sitcoms and movies usually mentioned menstruation in a brief, joking way, teaching euphemisms and stereotypes about menstrual behavior, magazines aimed at girls and women added personal experiences to their advice about menstruation contained in their health columns. *Seventeen* included stories about menstruation in its "trauma-rama" column, in which teenagers wrote in to tell tales of embarrassment and humiliation. In 1998, one teen wrote about her annoying relatives picking on her for her grouchy mood on Christmas day, concluding, "Finally, I just couldn't take it anymore. I burst into tears and screamed, 'I'm having my period, alright? Give it up!' The entire room went silent." Another told a story of a Christmas party that included a "cool guy cousin" who "was best friends with my crush" and an aunt who gave her a training bra "and loudly declared, 'Congratulations to Sue, who recently got her period. No more stuffing her bra for school.' I was mortified. How was I going to explain to my cousin that I did *not* stuff my bra!"[57] A story about the experience of being rear-ended narrated, "After an embarrassing glove-box search that turned up, among other things, a lifetime supply of Tampax and about 300 Bazooka comic strips, I forked over the necessary documents."[58] Tampons were a common theme; in another "Trauma-Rama" column, a girl and her date look in her purse for their movie tickets, and "my stash of tampons, which was right next to my wallet, was staring us in the face." She was horrified but reported, "Being the big sweetheart that he is, he said, 'It's OK.' And then, just to make me feel better, he added, 'I didn't see anything, anyway.'"[59]

All of these types of media aimed at teens and preteens gave the youngest women and men interviewed a tremendous amount of information about menstruation to supplement the scientific account offered in sex education classes and manufacturers' pamphlets. But even before they encountered a lot of published accounts of menstruation, many had learned something from their mothers at home. These mothers, from the generation of women born in the 1940s and 1950s, had particularly embraced a thorough, scientific education about menstruation, communicated in carefully carved-out spaces including school sex education classes, sit-down conversations between mothers and daughters, and private reading of pamphlets and books. Building on this base, some of these mothers broadened this acceptable space for talking about menstruation without embarrassment or shame, conversing more openly with their immediate family members, especially in the privacy of their homes. Tim Dai explained that his mother, a teacher, is "just very open about these things, so she, I think, really did want me and my sister to be aware of ourselves and the way people are . . . I just remember the notion of her openness and frank discussion and, yeah, I think she used words, like she would talk about "period" or whatever, and I just remember being aware of it in that sense.

Some young women remembered being in the bathroom with their mothers and seeing them menstruating. Stacy Joyce Lockert, the sister of Carlen Joyce Thomas and daughter of Dorothy Joyce, remembered being really little, going, "Mommy, why is, [laughter], what's that in the toilet? Why are you bleeding in the toilet?" It scared me. I don't remember explicitly how she explained it to me, but then I was ok with it . . . I remember seeing my mother, she was very open, her showering, and being in the bathroom, and doing her thing. Quizzical as I was, I would stare, and watch, and ask questions, so I knew all about it.

Unlike Carlen, Stacy experienced their mother as very open about menstruation. This may reflect the exigencies of parenting the youngest of three children, but it may also reflect the evolution of Dorothy's "modern" perspective over her lifetime.

It was mostly daughters who recalled seeing direct evidence of their mothers' menstruation, but at least some women in this generation were more open with sons, as well. Peggy Woo laughed as she told a story about getting dressed in front of her young son when she had her period.

When he was little—he was probably about five—and he's laying on the bed, and I was walking around. I had taken a shower, and I was walking around naked, and I had a tampon in so I had a string hanging out . . . I was walking around, and I could see him looking, as I was walking around my room getting dressed. I could see him looking, following the string with his eyes. Like, looking around, and I knew he was watching it. And I didn't say anything, I just thought, "OK, this is going to be interesting!" [Laughter.] And then he said, "Mommy," in that questioning tone, and I just said, "Yeah?" He goes, "Mommy, come here." And I said, "Why?" And he says, "Come here and let me see that rope." [Laughter.] He called it a rope! . . . I just said, "It's not a rope, it's a string." And I said, "No, I'm not coming over there." [Laughter.] "I have to get dressed." I mean, what are we going to do, play with it? So, I figured, at five, he doesn't need to know.

Peggy might not have explained it to her son, but she did not consider it something to hide from him. Her daughter got a bit more explanation by the time she was seven years old.

I've already told her, sort of. In a way she could sort of understand. Because she's very inquisitive. And she sees everything I do. So she's asking me about these tampons, and "What are you doing?" and all that kind of stuff. So I've made very sketchy explanations about bleeding every month, and that girls do, and she was saying, "Oh, I don't want to do that!" [laughter], and I said, "Well, you will when you're ready." She said, "Well, I don't want to now." I said, "You won't now. It's maybe when you're like, thirteen, or something like that," and I said, "Then you'll know all about it, don't worry." So she's had that talk.

This kind of openness, allowing children to be in the bathroom in the first place, and to see the blood itself as well as the way menstrual technology was actually used, was a radical change from the previous generation. This kind and level of openness about menstruation was not adopted by the majority of women interviewed, but it seems to have been increasingly thinkable and acceptable, at a time when parents became more willing to allow children to see them unclothed and engaging in bodily functions. The 1960s counterculture had promoted a cluster of ideas about the body, that it should be "natural," liberated through free pursuit of sexual pleasure, and unconstrained by shame and modesty. Those who participated in the counterculture understood these ideas to be in opposition to mainstream, middle-class culture. However, when it came to menstruation, women increased their openness about their bodies with their family members as part of becoming "modern" and joining a broad, self-perceived middle class. They expanded the territory of appropriate places where menstruation could be mentioned and appropriate ways for it to be observed, not in a way that broke with modern, middle-class culture but in a way that enhanced it. Young women who learned as children about menstruation from their mothers in the bathroom could handle it themselves with confidence and efficiency later, avoiding even more completely the anxiety their great-grandmothers experienced at menarche. Young men who learned from their mothers could be more appropriately supportive of their mothers, sisters, girlfriends, and wives, assisting women in their efforts to manage menstruation rather than accidentally undermining them.

A century of efforts by individual women and girls, sex educators, teachers, and pamphlet writers and distributors had transformed how women learned about menstruation. In the early decades of the century, women like Ida Smithson and Samantha Fried were shocked, scared, and even ashamed when they got their first periods. By the end of the century, almost all American girls heard a reassuringly phrased scientific account of menstruation before they reached menarche, as well as lots of supplemental information from mass media, and most boys learned something as well. Books, pamphlets, and school education programs helped parents and teachers to carve out safe, carefully defined venues and vocabularies for teaching about menstruation. This "modern" way of learning about menstruation gave Americans the confidence and knowledge they needed to handle menstruation in a modern way, efficiently, effectively, and without anxiety. They had enough information that they could choose to talk about menstruation if they wanted, certainly. But they were rarely forced to discuss it outside of the circumstances they regarded as appropriate because they were not caught by surprise or racked with anxiety. The women who allowed their daughters and sons to hear about and see evidence of menstrual management at the end of the century continued the logic of the "modern" way to learn about menstruation, expanding acceptable spaces of communication from classrooms, mother-daughter talks, and private pamphlet reading to conversations within the family and exchanges with children in the bathroom or bedroom. They maintained the expectation that learning about menstruation would help girls to handle it in a way that minimized its impact on their lives.

This transformation of how girls and boys learned about menstruation was only one aspect of the twentieth century creation of a "modern" menstruating body. Simultaneously, there came a related shift in how women, health care providers, physical education teachers, industrial hygienists, and popular health writers understood the relationship between health, discomfort, and menstruation.

CHAPTER THREE

The Modern Way to Behave while Menstruating

Changing Health Beliefs and Practices

Between the late nineteenth century and the middle of the twentieth century, medical and popular beliefs about the relationship between menstruation and general health shifted dramatically. By the 1990s, women and physicians in United States were remarkably unconcerned about menstruation as an indicator of general health or illness. Even more definitively, they stopped worrying about menstruation as a time of particular bodily vulnerability. Previously wide-ranging concerns about menstrual irregularity and pain were narrowed to a focus on relieving pain in order to continue normal work and play. Of course, menstruation did not completely cease to be an indicator of health or illness; too much disruption in a woman's usual pattern of menstruation caused concerns about specific problems with reproductive organs. But these occasional concerns about reproductive health were completely overshadowed by the new way of thinking about menstruation in terms of managing blood and discomfort so that they did not interfere with everyday life.

These changing ideas about health and menstruation allowed women and their employers, teachers, doctors, and sexual partners to construct a new set of expectations about which activities were appropriate during menstruation. The "modern" body they envisioned did not need any special attention or treatment during menstruation; women were expected to be able to play sports, work, and even have sex during their periods, and if these activities caused discomfort, they were supposed to call on their doctors to remedy the pain. These new expectations were not uncontroversial, and not all women felt obliged to meet all of them. Experts sometimes challenged them as well, concerned that the focus on maintaining women's productivity all month might endanger their reproductivity. Nonetheless, the push for women to continue normal activities during menstruation grew stronger and stronger over the course of the century, feeding into the larger vision of the well-managed, constantly productive modern body.

Indications of initiation of major change came in the last quarter of the nineteenth century, when the movement of "new women" into colleges and into public life more generally sparked an extended, published discussion of menstruation in the medical and popular press. Fierce debate pitted the most conservative medical writers, who saw menstruation as particularly debilitating and menstruating girls and women as uniquely fragile, against women reformers who made radical claims about women's capabilities in many areas. Although a few generations later, even those radical reformers would appear quaint and old-fashioned in their concern about protecting menstruating girls' health, their protests against conservative medical views marked the beginning of a "modern" way to think about menstrual health.

Conservative medical thinkers were led by Dr. Edward Clarke, a prominent faculty member at Harvard Medical School, who published an alarming book, *Sex in Education; or, a Fair Chance for the Girls* (1873). Clarke claimed that women who studied hard and competed with boys in preparatory school and college, ignoring their periods and their pubertal development, ruined their health and often were unable to have children. Clarke called upon a popular medical theory that the body had a limited supply of vital force. This vital force had to be carefully husbanded for maintenance of health. Men were often warned that masturbation and excessive coition would drain vital force from other necessary bodily activities and reduce mental abilities. Clarke applied the medical theory to girls at puberty, arguing that the vital force necessary to grow their reproductive organs and establish their functional regularity would be sucked up during intensive study, resulting in general and reproductive feebleness, debility, and even death. This damage could not be immediately detected; a girl might feel perfectly healthy during her studies, only finding later that she had damaged her reproductive system beyond repair. Or, a girl might suffer profoundly during her periods, a sign that she was damaging herself, but hide it from her teachers in her ambition to compete with the boys. He recommended that girls rest entirely during their periods and take an easy course of study during adolescence.[1]

Clarke made his recommendations, he said, both to benefit individual girls and to protect "the race." Like many of his time, Clarke feared that white Anglo-Saxons were soon to be outnumbered by those he considered to be "racially" inferior, and he made eugenicist arguments promoting childbearing among white, middle-class, Anglo-Americans. Clarke's medical and social concerns, and his recommendations to women, built upon a longer history of Victorian understandings of white women's health and reproductive role. Middle-class white women were widely understood to be delicate, at the mercy of their reproductive organs, and becoming more so by the decade.[2] Each generation of women bore fewer children, and many women may have adopted a semi-invalid identity in order to avoid constant childbearing, reinforcing medical perceptions of their frailty.[3] Many who commented with concern upon these trends believed that women were inherently delicate and were further harmed by what medical writers saw as the "artificial" style of city living adopted by a growing segment of the population, in contrast to "natural" rural life. As poor Irish Catholics began to arrive in large numbers in the 1840s and 1850s, and Eastern and Southern Europeans in the 1880s and 1890s, anxiety about white, Anglo-Saxon, Protestant women's reduced childbearing was figured in eugenic terms, as a national and race problem as much as a personal one.[4] Middle-class white women who were increasingly choosing to have fewer children, more years of education, and more political and social influence were accused by Clarke and others of ruining their own health and morals, and the nation's.

Response to Clarke's book was immediate and strongly divided. Many physicians supported him; however, the "new women" he criticized for their ambitions beyond childbearing were vociferous and eloquent in their attack on his book. A year after it was published, women's suffrage movement leader Julia Ward Howe gathered the flurry of critical editorials and reports it had provoked into a book, *Sex and Education: A Reply to Dr. E. H. Clarke's "Sex in Education."* Prominent reformers, representatives of women's colleges, women's clubs, and anonymous editorialists joined in defending women's collegiate education.

These defenders of women's education first criticized Clarke's science. Howe protested that Clarke's book was a polemic masquerading as science, with broad generalizations drawn through poor reasoning from only a few medical cases. Elizabeth Stuart Phelps doubted that those cases were representative of healthy women. "The physician knows sick women almost only. Well women keep away from him, and thank Heaven. If there be any well women he is always in doubt." Clarke's critics found his blaming of higher education for girls' poor health extremely suspect. Maria A. Elmore pointed out that Clarke had no complaint about women working hard all month at home, in factories, and in retail shops, as long as they were following traditionally female vocations. She and others believed that Clarke was making a medical argument purely in service of a political attempt to keep women out of higher education and traditionally male professions. Clarke did indeed claim that brain effort was much more taxing than physical effort and that housework was much safer for adolescent girls than serious study. A few of Clarke's contemporaries, such as Azel Ames, were also concerned about the effects of industrial work on adolescent girls' development, but, for the most part, Elmore was correct that medical arguments about girls' fragility were directed at ambitious middle-class girls and women.[5]

Howe and others believed that, contrary to Clarke's claims, schools were on average improving girls' health by educating them about how to care for their bodies. Howe and others did not dispute Clarke's eugenic arguments about needing to preserve the race; rather, they claimed that education for women promoted white middle-class women's health and reproduction rather than undermining it. Representatives of six women's and coeducational colleges testified that their women students improved rather than ruined their health during college. It was true that large numbers of college-educated women from this generation did not marry or have children, but it was not due to health deficiencies. While many editorialists acknowledged that some secondary schools pushed girls to study too hard, they believed that reasonable, hygienic guidelines for study needed to be set for girls *and* boys. Caroline D. Hall suggested the problem was the work girls did in addition to their studies, not the studies themselves. "If boys are preparing for college, they do not have to take care of the baby, make the beds, or help to serve the meals."[6]

Howe and the other editorialists had a number of alternate suggestions why a girl's or woman's health might break down. Howe suggested that it was because "girls have the dispiriting prospect of a secondary and derivative existence, with only so much room allowed them as may not cramp the full sweep of the other sex." Mrs. Horace Mann believed that "unfortunate marriages are the circumstances under which the harmonious development of nature is arrested and perverted." Elizabeth Stuart Phelps noted that in all of Clarke's cases, the girl's health broke down *after* leaving school. She suggested that a girl could be "made an invalid by exchanging the wholesome pursuit of sufficient and worthy aims for the unrelieved routine of a dependent domestic life, from which all aim has departed, or for the whirl of false excitements and falser contents which she calls society." Clearly, if Clarke's critics were correct, extended schooling was better for women than more traditional options.[7]

One of the most influential critiques of Clarke was by Mary Putnam Jacobi, a prominent physician and researcher who won Harvard's Boylston prize for her lengthy essay "The Question of Rest for Women during Menstruation" in 1876. Jacobi approached the question of whether women needed rest during menstruation from theoretical, experimental, and epidemiological perspectives. She theorized that the menstrual cycle was a "nutritive wave," in which a woman's body continuously produced nutritional surplus to support reproduction and cast it off in the form of menstruation when it was not needed. In pointing out the ways in which menstruation was most appropriately compared to digestion and excretion, she argued that menstruation was no more taxing on a woman's body than these other processes. She conducted experiments on a small group of women to show evidence of a nutritive wave in all of a woman's systems, in ways that allowed women's bodies to lose blood each month without being weakened. Finally, she distributed a questionnaire, which was completed by 268 women, to find out whether women experienced mental or physical difficulties during menstruation, or had health problems that might be the long-term ill effects of regular activity during menstruation. While nearly half of Jacobi's interviewees experienced serious pain during menstruation at some time during their lives, she concluded that it could not be blamed on education, since, of the women she surveyed, those with more education seemed to suffer less. She argued from her data that a poor constitution, either inherited or developed in childhood from lack of daily outdoor exercise, accounted for much of the menstrual pain her interviewees experienced.[8]

Advice about menstruation in popular health books reflected the opposing perspectives of Clarke and Jacobi from the 1880s through the first two decades of the twentieth century. Health reformer John Harvey Kellogg recommended in 1891 that as soon as a girl experienced premenstrual symptoms until the end of her period, she "should be relieved of taxing duties of every description, and should be allowed to yield herself to the feeling of *malaise*, which usually comes over her at this period, lounging on the sofa or using her time as she pleases." In 1898 Henry Lyman and his coauthors suggested that girls be taken out of school during menstrual periods, and even for all of puberty if a girl was overly ambitious and inclined to study hard. In contrast, in 1885 Alice Bunker Stockham noted that many women felt a need for some extra rest during menstruation, but not everyone: "On the contrary some have found that congestion and pain are relieved by occupation sufficient to interest the mind, with exercise adapted to increase the circulation." She believed that all who felt up to working during menstruation should do so and that many more women would feel up to it if they loosened their corsets. "The fact is that girls and women can bear study, but they can not bear compressed viscera, tortured stomachs and a misplaced uterus."[9]

Despite their strident and politically potent disagreements, health writers from both sides of the debate continued to share many concerns and assumptions into the early twentieth century. All maintained the millennia-old belief that girls and women should avoid cold water and getting chilled during menstruation, because the flow might be checked, causing internal "disorder" and inflammation, and because women were more susceptible during menstruation to catching colds that could lead to long-term health problems. They also maintained the traditional belief that mental shock could suppress menstruation, again leading to health problems of all kinds.

Writers on both sides of the debate also shared widely held concerns that as America industrialized and urbanized, its people were becoming weaker and sicker, and passing their weaknesses from generation to generation. Jacobi and Stockham, like Clarke and Kellogg, believed that girls were less healthy than they could be, in ways not entirely redeemable through personal reform. Jacobi pointed to the inheritance of weak health from parents and from improper practices during childhood. Stockham believed that *"a woman in perfect health* need take no especial care and make no change in her manner of life at this period. But under our artificial habits of life, such a woman is the exception rather than the rule, and in most cases some attention must be paid to the recurrence of the menses."[10] Nearly all health writers looked nostalgically to what they believed was a healthier rural past and agreed that "artificial" practices such as wearing tight corsets, staying inside all day, and expecting girls to be "ladylike" rather than running and playing were detrimental to girls' health and development, often permanently. Jacobi and Stockham, however, were much more optimistic that reforming general health habits could at least improve girls' and women's menstrual health.

—⁂—

Jacobi and Stockham's perspective, daring and progressive in the late nineteenth century, would soon seem quaint and old-fashioned. Their assertions of women's ongoing health even during menstrual periods broke the ground for a new way of understanding menstruation, which would soon mark a more radical break with Clarke and Kellogg, as well as with centuries of medical and popular belief. While advice literature maintained a conservative tone though the 1910s, physical educators and physicians at women's and coed colleges were developing a new perspective and a new set of practices. By the 1920s, they would begin to insist that practically all girls could menstruate painlessly and develop healthfully while attending school and playing sports if they only adopted correct practices, even as late as college age.

Since at least the late eighteenth century, health advice authors such as William Buchan had blamed tight corsets, poor diets, and restrictions on outdoor physical play for a large proportion of girls' troubles with menstruation. At the beginning of the twentieth century, college physical educators found themselves in a position where they could enforce better "hygiene" habits, a term that then encompassed all the daily practices that impacted a person's health. Colleges could require students to maintain regular sleep patterns, provide meals considered to be balanced and healthy, and require regular exercise such as daily outdoor walks and calisthenics at the gymnasium. Helped along by changes in fashion, they could even coax or coerce college women into wearing less restrictive corsets and lighter skirts, especially while they exercised. None of these ideas was new, but the circumstances whereby these hygienic practices could be enforced on relatively large groups of young women were novel. And in single-sex college settings, at least, it appears that women enthusiastically embraced these opportunities to eat heartily and take long walks or play sports in the relative freedom of "bloomers" and other clothing that was lighter and less restrictive than what they usually wore.[11]

When it came to menstruation, physical education and hygiene departments crafted specific policies designed to protect women's health and well-being without allowing any unnecessary coddling. As historian Martha Verbrugge has shown, they first redefined the scope of concern about women's periods. As the science behind Clarke's theory of limited vital force fell into disrepute, physical educators ignored his concerns about the impact on the entire body of exercise during menstruation and focused on the uterus itself. They argued in their professional literature that women could and should continue their normal activities, including moderate exercise, during menstruation; these activities kept the body strong and the blood circulating and thereby relieved menstrual pain. The only activities they regarded as problematic were those that jarred the uterus and burdened the uterine ligaments when the uterus was heavy with menstrual blood.[12]

Concern about uterine ligaments stretching during menstruation and causing uterine prolapse was not initiated by physical educators, but they elevated this concern above all others and magnified it. Before the nineteenth century, medical advice writers seemed to assume that women would do their usual work during menstruation, even heavy work, and only advised against it if it was causing profuse menstruation leading to weakness and prostration. In the nineteenth century, a few medical writers expressed concern about the possibility of uterine prolapse but certainly did not emphasize it. It might seem that in focusing on yet another possible problem with women's bodies during menstruation, physical educators were simply expanding the grounds for claims of women's general debility. However, in creating this new focus for concern, physical educators demonstrated that they were concerned about their students' health and aware of potential issues, while creating narrowly defined restrictions on women's behavior. In telling students not to play competitive sports during menstruation, they implicitly permitted all gentler activities during menstruation and competitive athletics the rest of the month. These were much less onerous restrictions than those suggested by Clarke and his ilk. This focus on uterine anatomy also allowed physical educators to sidestep or ignore traditional concerns about cold water, chills, and emotional shock.

As physical educators grew increasingly confident that blood-filled uteri could handle some bouncing and jouncing, they gradually increased the physical education requirements for women students during menstruation. For example, by 1921, Smith College began to require women to attend physical education classes during their periods, though they could participate in alternate, gentler activities. Notably, this timing coincides with the mass marketing of Kotex, as well as increasingly liberal attitudes toward exercise during menstruation. By the late 1920s, Smith women were participating in most normal gym activities; by 1936, the year Tampax was introduced, horseback riding, an activity previously forbidden during menstruation, was allowed and only swimming remained off limits. Not every college moved in this direction as quickly as Smith, but the trend was evident in many institutions.[13]

Women physicians associated with women's and coed colleges conducted research that often supported and complemented physical educators' efforts. Like the physical educators, they were working with a healthy population, not sick women patients. They also were optimistic that women could be healthy and comfortable all month, with few limits on their activities. Clelia Duel Mosher, a professor of personal hygiene and physician for women students at Stanford University in the early twentieth century, recorded and analyzed hundreds of her students' experiences with menstruation over a number of years. In comparing the data she had collected in 1894 with that collected in 1915–16, she noted that as fashions had changed, with corsets broadening and the weight of skirts diminishing, the number of students reporting menstrual pain diminished drastically. In 1894 only 19 percent of the women she studied were free from pain, while by 1916 that number had reached 68 percent. Mosher was sure that other reforms that complemented the change in dress would take care of most of the rest of the reported menstrual pain. She had demonstrated success with a series of abdominal strengthening exercises, which were widely recommended in advice books and pamphlets through the 1950s. In 1923, she declared that the "modern college woman" tended to have trouble remembering the details of her periods because she had such a healthy attitude toward them. "This periodic function of menstruation makes no more impression on her mind than does that other periodic function, digestion." Mosher believed that even the pain that was reported was an overreporting of trouble, since women who took painless menstruation for granted would remember and report exceptional troublesome cycles.[14]

Psychologist Leta Stetter Hollingsworth took on the issue of capability during menstruation in her Columbia doctoral dissertation, measuring a group of women's mental and motor skills over the course of a month. She found no variability, leading her to conclude that assumptions about women's menstrual incapacity were unwarranted.[15] By the 1920s, women researchers and physical educators had gathered a great deal of scientific evidence and practical experience demonstrating that women should not expect to be incapacitated during their periods, either because they were less capable or because they would injure themselves by continuing normal activities. At a time when college students were becoming closely watched and widely imitated trendsetters in American culture, changes in college students' handling of menstruation were likely influential beyond the campus gates.[16]

Colleges were not the only places where the relationship between menstruation and young women's activities was newly addressed. Young white women, with and without college degrees, increasingly worked outside the home. With the rise of corporate and government bureaucracies, and the blooming of new retail outlets and a cultural emphasis on consumption, many moderately paid office and sales jobs were created, blurring the nineteenth-century blue-collar/white-collar class line. Many of these "pink collar" positions were filled by unmarried white women. For example, Mary Hanson, like many young women, graduated from high school and joined the telephone company as an operator in the early 1920s, at a time when connections between parties had to be made by hand. Eight years later, when she married, she was fired because the telephone company had a policy against employing married women, but she found similar work as a switchboard operator at a department store. Nursing and teaching also offered professional opportunities for more-educated women as the fields expanded and feminized.[17]

Women's work had implications for their families' class status in two ways. First, for many families the wages that women earned were what allowed the family to be able to purchase the consumer goods that made life feel "comfortable," middle class rather than struggling. As Mary Hanson put it, with her income in addition to her husband's earnings as a firefighter, "We weren't rich by any means, and we weren't poor. We had enough to eat, and we managed to have a nice living." Second, like Mary Hanson, many women who worked in pink-collar jobs had husbands who worked in blue-collar jobs, so the wife, in her public persona on the job, most clearly represented the family's middle-class or aspiring middle-class status. African American women did not have ready access to clerical and sales jobs outside of African American enclaves until the 1960s, but once they gained wider access to such jobs, their families often followed the same pattern. Anika Taylor's father, a truck driver, was the primary earner in her family, but her mother also worked at an insurance company during the 1980s and 1990s, until her fourth child was born. Earlier in the century, African American families in the South were more likely to send daughters than sons to school. In many cases these women had greater access to white collar employment, especially as teachers in the black community, than their African American brothers and husbands. For aspiring white and black families, then, women's work outside the home was important in terms of finances and self-representation.[18]

As industry expanded and more women came to work in large offices, department stores, and factories, many of the biggest companies opened departments of industrial hygiene and health.[19] They hired doctors and nurses whose job it was to keep employees well and productive, help sick workers to regain health and productivity as quickly as possible, and screen out applicants who had health conditions that would make them less productive. These industrial health care workers also conducted studies of their patients and treatment plans and published the results in industrial hygiene journals. While physical educators addressed the bodies and concerns of college students, industrial health workers expanded their research to include those working in offices and department stores, the aspiring rather than the established middle class.

Industrial physicians and nurses took many cues from physical educators in their understanding and treatment of menstruation, though they had quite different goals. In the 1920s, a number of studies focused on menstrual pain as a cause of loss of productivity, in contrast to the nineteenth-century studies, which focused on the bad effects of work on menstruation and, in turn, on general health. Several studies concluded that dysmenorrhea did, indeed, cause some loss of productivity, but this difficulty could be largely remedied by the actions of a conscientious health department, and most women could be made to work at normal capacity all month. A number of factory and department store health departments gave women advice about keeping their bowels clear and maintaining good sleeping and eating habits; provided those in pain with cots, hot water bottles, hot drinks, and occasionally, drugs; and referred them to physicians when their pain persisted despite these efforts to address it. The cots, water bottles, and hot drinks provided ways to address menstruation at work that were similar to what older women interviewed for this book who suffered from severe menstrual pain did for themselves at home.

The Metropolitan Life Insurance Company, an innovator in the large-scale hiring of women clerks and in the spatial and social organization of gender in the office, had a particularly progressive and research-oriented health department. The company attended to its young women workers' social well-being by giving them their own office and rest room space—thus separating them from extensive or intimate contact with male employees and customers—and to their physical well-being by offering those who suffered from persistent dysmenorrhea an exercise program to alleviate it.[20] In fact, the company did more than offer the program; in 1928, they insisted that 523 young women who had reported to the rest rooms repeatedly because of menstrual discomfort participate in an experimental daily exercise regimen. Each woman's menstrual history was recorded by a physician, she was made to undress and have her posture traced in outline on a medical chart, and she was educated about "the necessity for regular bowel habits, proper diet, and fluid intake." She was shown Mosher's abdominal contraction exercises, and sit-ups were added as well, after a few months, on the recommendation of another feminist physician. Each woman was required to report weekly for a check of her progress in exercises and posture. Women whose dysmenorrhea did not improve quickly were given a pelvic examination; those who appeared abnormal were referred to private physicians, while those who appeared normal were required to report daily for group exercise classes. The researcher, Dr. Ruth Ewing, reported a tremendous success rate of improvement in 81 percent of the cases and attributed most of the residual problems to some women's lack of cooperation. She also noted that the program had a positive effect on the productivity of women who did not participate in the research as well, because "employees whose dysmenorrhea was slight were less apt to stay away from work when they knew that their absences would be noted and their history carefully checked." The exercises probably really were helpful for a lot of women—90 percent of those who participated in the research replied in an anonymous survey that they had benefited from the treatment—but the high level of surveillance and the publicized assumption that if women had menstrual pain, it was the result of their own laziness, probably also had a lot to do with the program's success.[21]

Like physical educators, company physicians were optimistic that most women could be cured of pain if they were willing to obey the rules of correct "menstrual hygiene." They used *hygiene* in the broad sense in which it was often used in the nineteenth and early twentieth centuries, to mean all habits or practices that enhanced or protected health. They did research to demonstrate the efficacy of correct menstrual hygiene, and their professional journals abstracted articles from the medical literature, demonstrating in lab experiments that there was no drop in muscular and mental efficiency during menstruation.[22] Some researchers, including Dr. Ewing of Met Life, insisted optimistically that everyone with normal reproductive organs could be cured of their dysmenorrhea. Others, such as Dr. Margaret Castex Sturgis of the Women's Medical College of Pennsylvania, believed that while many could be cured of their dysmenorrhea through either the efforts of a company health department or a private physician, there were a few who would never be cured, and they should be treated as special cases. "The woman employees who must remain at home every period because of severe disability should be considered in the industrial world as any other handicapped person. The question whether they should be employed is a debatable one, but their disability should not be charged against the great majority of working women among whom lowered efficiency as a result of menstruation is negligible." In fact, in the department store Dr. Sturgis studied, women who took absences to deal with their periods were fired. Dr. Sturgis and others were willing to shut a few women out of the workforce because of menstrual problems, as long as the rest of the women were understood to be as capable as men throughout the month.[23]

Not everyone, however, was out to prove that women were efficient workers or could be made so with appropriate industrial hygiene measures. In 1920, an Ohio rubber company asked its personnel research director to assess whether female office workers deserved the same salary as male office workers. The research director noted that women lost 6.02 percent of their available work time to absences, while men lost only 2.42 percent. He investigated the reason for these absences in a strikingly intrusive fashion. "The company employed visiting nurses who called at the home of the absentee on the same day the absence occurred. The visiting nurse ascertained and reported the real cause of the absence, in so far as possible, and the nurse's statement of the cause was accepted rather than the statement of the employee." Since he found that 17 percent of women's absences were caused by "ailments peculiar to women only: dysmenorrhea and ovarian congestion," he concluded that women, though they "may be *intellectually* competent to undertake any and all vocations . . . do not deserve the same pay as men even when they hold the same kinds of positions." It seems likely that women were docked more than the 1.02 percent of pay they presumably should have forfeited collectively for absences caused by female problems, or even the

3.6 percent for their total increased absences. Given this kind of argument, it is clear why other researchers were so concerned to differentiate between "normal" women, who were just as efficient as men, and menstrually "disabled" women, who were a very small group that it was reasonable to treat differently.[24]

———※———

Despite the few ever-present naysayers, the general consensus that most women were healthy and physically and mentally able during menstruation was amplified and widely publicized in menstrual product manufacturers' advice pamphlets. Kimberly-Clark's first educational pamphlet, written by nurse Ellen Buckland in 1923, was blithely optimistic about the level of menstrual health and comfort women could attain through the use of Kotex, plus a few other simple hygienic practices. In *Now Science Helps Women Solve an Age-Old Problem*, Buckland told women to stay active, take long, brisk walks, drink plenty of water to prevent constipation, maintain an upbeat attitude and positive expectations, and avoid infection by rejecting cloth pads and choosing disposable Kotex instead. She cited research from women's colleges in support of her recommendations and simply ignored more controversial topics, such as whether competitive sports were acceptable. This first Kotex pamphlet was written in a genre somewhere in between old-time patent medicine promotion booklets, which made fantastic, impossible claims about nostrums returning women from the brink of death to perfect health, and modern corporate-sponsored educational pamphlets intended to appear sober, authoritative, and even-handed. Buckland held out the hope that using the product she promoted, plus a few other health practices, would make menstrual difficulties and annoyances disappear from one's life forever. She concluded, "I hope that what I have told you here will help you discard the outworn superstition that women are handicapped by nature and must accept mental, emotional and physical derangement once every month. Instead of being natural, this is unnatural, as hundreds of thousands of happy, well-poised, efficient modern women know." Buckland, and the corporation she represented, were out to define the "modern" woman as a "happy, well-poised, efficient" Kotex user. This modern woman was the same woman whom physical education teachers believed they were training in their classrooms, and whom industrial hygienists wanted to hire as a corporate worker. Kotex pamphlets drew upon and reinforced the model of the modern menstruating woman as it was being developed on college campuses and in corporate life.[25]

Buckland also included advice to mothers to think carefully about how they told their daughters about menstruation and to give girls "practical, efficient, up-to-date Kotex" so they could avoid "the nervous dread" that women felt when they had to worry that their menstrual blood or cloth pads would show. She warned, "Don't let her inherit the superstitions and suffer from the inconveniences which have afflicted you." Since the late eighteenth century, advice manuals had urged mothers to talk to their daughters about menstruation, but they had focused on sharing information about proper management of menstruation and preventing girls' shock at menarche. Buckland's advice, novel in the 1910s, became common in health manuals and educational materials in the 1920s. Mothers were urged to demonstrate a "modern" attitude toward menstruation, if not for their own sake, then at least for the sake of their daughters. Health writers became more and more convinced that a great deal of girls' menstrual suffering was psychosomatic, attributable to the expectation that menstruation would be painful and difficult. Girls who saw their mothers complain and moan and lie in bed during menstruation, the theory went, would dwell upon every twinge and cramp, and their anxiety would magnify their pain.

As Kimberly-Clark began to produce pamphlets intended for girls rather than adult women, the tone and content of the advice shifted slightly. Adopting the role of the socially responsible corporation, Kimberly-Clark produced pamphlets that gave more balanced and educational information about menstruation, and at least created the appearance that the primary goal was education rather than selling Kotex. The first widely distributed pamphlet for girls, *Marjorie May's Twelfth Birthday*, focused in its first edition on giving a brief explanation of menstruation and the use of Kotex to manage it, but by the 1932 edition, it included some basic health advice. First, it labeled abstention from bathing during menstruation, common practice through the early twentieth century, as a "harmful superstition." It assured girls that they could and should bathe during menstruation as long as they took "a *warm* soap bath, but never a cold, and never a hot bath."[26] This advice reflected major changes in American health beliefs. Through the middle of the nineteenth century, bathing was a suspect practice in general, not just during menstruation. Before indoor plumbing became common—not until the late nineteenth century in urban areas and later in rural areas—indoor bathing required laborious preparation, and outdoor bathing involved cold water and exposure to cold air. Doctors and lay people alike worried that this could lead to chilling, which could lead to a cold, which could turn into a more threatening disease. Women were thought to be particularly vulnerable to chilling and illness during menstruation, so even when regular bathing became more common, women would generally abstain during menstruation.

It was not only indoor plumbing and the eventual introduction of easier ways to heat water that led to greater popularity of regular bathing; at least as important was the rise of a middle-class culture of respectability, which adopted cleanliness of the body as one of its central tenets. Advice literature for the middle class began recommending daily bathing by the 1840s, and in the antebellum period, regular bathing, among other practices, distinguished a small American middle class from the masses. In the late nineteenth century, advice writers for women began to sanction bathing during menstruation, done cautiously, despite the health risks, because menstrual odor was felt to undermine respectability. They distinguished the minor risks from cautious bathing in heated water from the much more significant risk of becoming chilled due to exposure to the elements. Into the early twentieth century, concern about exposure to cold and chilling remained high, but bathing in warm water gained full acceptance in the advice literature.[27]

During the Progressive era, rather than maintaining regular bathing as a mark of class distinction, middle-class reformers worked hard to convince everyone, particularly the urban immigrant poor, to take daily baths. Anxious about the diseases they feared might be harbored by dirty bodies and repulsed by those whose bodily practices differed dramatically from their own, reformers tried to remake the urban poor in their own images. When it came to daily bathing, for the most part they succeeded. Some of the practices taught by Progressive reformers were critical to good health in an urban setting; others promoted "American" self-presentation rather than health. Many immigrants and their children, like Samantha Fried, were happy to participate in Americanization projects. Even those who were not eager to abandon old world customs generally realized it was important to their economic success to assimilate as the middle-class reformers urged. Black and white settlement workers and activists made similar reform efforts among poor African Americans. By the middle of the century, all but the poorest Americans had indoor plumbing, and regular bathing no longer distinguished a small, "respectable" middle class from the rest but instead separated the middle-class masses from a small impoverished class.[28]

In Kimberly-Clark's 1940 pamphlet, *As One Girl to Another*, warnings about exposure to cold water abounded but without nearly the alarm expressed a few decades earlier. Joseph Brown and Eli Greer, in their 1902 advice book *The New Tokology*, had warned that "chill is often caused by dabblings in water, and repression of the flow results, which paves the way to a future of invalidism." In contrast, *As One Girl to Another* offered no such dire predictions about the result of menstrual chilling. It told girls not to dive into cold water because it "gives your system a shock," to keep their feet dry to "avoid catching cold—at this time *espe-*

cially!" and warned that "getting chilled may cause trouble. So *don't* tumble about in the snow!" It seemed to anticipate that girls would expect to be able to do all kinds of things their foremothers probably would not have considered and warned them against each one. At the same time, none of the warnings came with much explanation or the suggestion that menstrual indiscretion would be life threatening. While advice literature continued to recommend avoiding cold water and chilling during menstruation into the late 1960s, over the course of the twentieth century it gradually lost its tone of urgency and threat.[29]

This pattern is reflected in the stories of those interviewed for this book. Among those born early in the century, many believed that they should refrain from swimming during their periods because they could catch pneumonia or the menstrual flow would stop. In contrast, women born in the 1940s and 1950s had often heard that they should not swim during menstruation, but few understood that this advice had been given out of concern for their health. Susan Ozols, daughter of a white New England quarry worker and a homemaker, thought that perhaps her grandmother thought she would get cramps if she swam during menstruation, but Susan never really asked; her husband joked, "You don't want to screw up the water!" Deborah Leary, daughter of a Boston upholsterer and a homemaker, explained that she was told not to go swimming "because you have to wear a pad. Can't wear one in a bathing suit, it's too obvious." Like many other women of her generation, once she started using tampons, she took for granted that she could swim during her period. Misunderstanding their mothers' advice, women of this generation assumed that once they had solved the sanitary problem of managing menstrual blood while wearing a bathing suit and immersing themselves in water, swimming during menstruation was a perfectly reasonable activity.

The fading of concern about exposure to cold during menstruation marked a radical shift in medical and popular understanding about menstruation and bodily vulnerability. A persistent concern for two millennia of Western medical tradition faded almost completely within the span of about fifty years. Even more sudden was the disappearance of concern about mental shock suppressing menstruation. Since at least the seventeenth century, doctors and women alike had worried that a sudden fright or other mental shock could suppress menstruation and thereby cause illness. Warnings of this sort appeared frequently in advice manuals through the end of the nineteenth century but abruptly disappeared almost completely after the turn of the twentieth century.

Many factors likely contributed to the rapid diminishment of ancient concerns about menstrual health. Worries about exposure to cold and to fright had originally been based on a plethora model of menstruation, within a humoral

understanding of the body. In this model, menstrual blood needed to flow regularly in order to keep a woman's body healthy. If menstruation was suppressed by cold or mental shock, the blood might stagnate within the body and cause a variety of illnesses. As physicians moved away from a humoral understanding of the body during the nineteenth century, they remained concerned about menstrual suppression, but there was little consistent medical reasoning about exactly why it was a problem. Doctors' concern appeared to be based more in the politics of women's education and participation in the professions than in a coherent medical theory, and once the political issues faded in importance, the medical

Opposite and *above*, menstrual dos and don'ts from "As One Girl to Another," an educational pamphlet for girls published by Kimberly-Clark, 1940 (courtesy of Duke University Rare Book, Manuscript and Special Collections Library, Durham, North Carolina)

reasons given no longer necessarily seemed persuasive. Given this lack of coherent or convincing medical justification for avoiding cold or mental shock, it was easy for expert advice and popular practice to shift in a direction commensurate with other aspects of the "modern" approach to menstruation. If the modern menstruating woman could work, study, and be physically active all month, she could probably bathe and swim as well.

In addition to advice about bathing and avoidance of cold water during menstruation, the 1932 *Marjorie May* gave the more recently popular advice about exercise promulgated by physical educators. "Some exercise, such as brisk walking and setting up exercises, are good for girls during the menstrual period, but doctors think it is harmful, during this time, to play hard games where there is much excitement and competition."[30] Through the early 1940s, Kotex pamphlets paralleled physical educators' writings warning girls not to "jiggle" their uteri too much during menstruation with activities involving jumping. Much of this advice persisted in manufacturers' pamphlets until the late 1960s, long after physical educators had stopped worrying about the delicacy of uterine ligaments during menstruation.

Advice from physical educators and physicians rarely noted the practical, technological problems of menstruation, although they were certainly highlighted in manufacturers' pamphlets, which promised that Kotex, Modess, or Tampax could solve most any menstrual discomfort. Older interviewees often limited their activities during menstruation, not because they were concerned for their health or had cramps but because the technologies they were using caused discomfort if they went about their normal activities. Mary Hanson would not have considered participating in a physical education class while she was using cloth "diapers," even if her school had offered one, because "it was the walking that would bother you. You'd get all chafed after a while. Yes, they're very uncomfortable." Rachel Cohen and her friends had the same problem with disposable pads, though she found ways to cope with it and be active anyway.

> Sanitary pads were very abrasive, especially the early ones that had this kind of gauze. They scraped your legs, and you got abrasions on your thigh, on the inside of your thigh. And so they called this abrasion soreness, too. So the tighter you pulled them, the less likely they were to scrape against your legs. And so I wore mine very tight. Because I was also very active. I mean, I played in school, I played in phys. ed., and so I ran around, and I didn't want that thing hanging down. So I used to make it very, very tight. But I remember almost all girls used to have big red abrasions about this size on each side of their thigh [indicates a circle with thumb and index finger], if they wore their pads loose, you'd have it on each side of your thigh. And it would take 'til the next month to heal, and then you'd start all over again.

Rachel tried tampons, but they were also less than ideal for her. "It would [laughter] slither out when you were running, and, you know, it was awful, you'd feel it, 'Oop, there it comes!' It was a horrible mess." Others had better experiences with tampons. One medical researcher investigating the risks and benefits of tampons favored tampon use partially because "a large number of patients [who tried tampons] have stated, without being asked, that pre-menstrual and menstrual tension has disappeared. This is partly because they are no longer bothered by an abdominal girdle and a band between the thighs. Many say they can forget that they are menstruating and so are without the disturbing annoyance they had every time they menstruated."[31] Technologically produced discomforts could be as disturbing to women at work and school as menstrual cramps were, but educators and industrial hygienists for the most part left it to menstrual products manufacturers to recognize the problem and recommend solutions.[32]

Manufacturers' pamphlets and advice books made it clear that even when the technology worked less than perfectly, and strategies for combating menstrual cramps were also not completely successful, it was no excuse for letting menstruation affect one's personality. An early example was Emma Walker's *Beauty through Hygiene: Common Sense Ways to Health for Girls* (1905). Walker advised, "Watch yourself carefully during these periods and if you observe that you are unusually irritable, keep a tighter grasp on your self-control and try to appreciate the fact that you are not quite your best self."[33] This type of advice increased in frequency over the next fifty years and made its way routinely into manufacturers' pamphlets. Modess's 1949 pamphlet advised the reader to "go out of your way not to be a 'sour-puss.'"[34] This pamphlet at least acknowledged that it could take a certain amount of effort to be cheerful during menstruation. Kimberly-Clark's 1952 *You're a Young Lady Now* was critical of girls who "cry easily, lose their tempers over nothing, and use menstruation as an excuse for being rude and mean." It claimed that "many girls *imagine* they feel worse than they actually do."[35] While most of the literature and technology related to menstruation was aimed at actually reducing discomfort and making menstruation feel less burdensome to girls and women, these bits of advice made it clear that menstruating in the "modern" way meant, at least by some experts' definitions, *appearing* not to be inconvenienced by menstruation as much as actually *feeling* unbothered by it.

While interviewees were generally enthusiastic about the "modern" way to menstruate, even those who were most enamored of this approach were not entirely willing to pretend to feel great when they actually did not. Women born in the 1940s and 1950s were eager to be "modern" about how they thought about and managed menstruation, but at least in certain circumstances, they interpreted being "modern" as having permission to talk about menstrual discomfort rather than being mandated to ignore or hide it. Most interviewees born before

1940 remained constrained by a previous standard of discretion, which encouraged women to hide menstruation from everyone, including male family members. They said that they would mention menstruation to their husbands to explain why it was not a good time to have sex but otherwise would not discuss it with them. Women born in the 1940s and 1950s, in contrast, were much more likely to point out a variety of other situations in which they would talk about menstruation with their husbands. Taking for granted their "companionate" marriages, women expected their husbands to be understanding confidants, and their husbands seemed to share that expectation. Barbara Ricci noted that "one reason it comes up is because sometimes you really are in a different mood, or you don't want to do certain things, and it may come up." Deborah Leary said, "Sometimes, the day before I'd get really bitchy," and she would tell her husband that her period was coming, "so he'd know. So he'd just kind of stay out of my way, wouldn't aggravate me." These women continued the openness they had established by participating in a particular model of menstrual education, broadening the range of acceptable venues and topics for discussion, and slowly transforming what it meant to be appropriately discreet about menstruation.

Menstruation advice literature, whether written by academic sex educators or employees of menstrual products companies, almost always told girls and women that if they had unmanageable menstrual discomfort or other issues, they should go straight to the doctor, who would surely be able to help them. In fact, women who sought their doctors' advice before 1960 had little chance of obtaining relief. Among those interviewed for this book, women who had especially severe menstrual cramps did seek help from physicians, but none were impressed with the treatments they were offered. Mimi Cummings Wood, Dorothy Joyce, Rachel Cohen, and Christina Donvito all went to physicians specifically to ask for help with menstrual pain, and none of them received very effective advice or medications. Mimi went as a teenager, but "other than to recommend Midol or something like that, he really didn't have an answer to my problem . . . No physician was able to tell me what the problem really was." Dorothy was not persuaded that her physician even took her problem seriously. "See those days, you went to a GYN man, and he was mostly taking care of pregnant women . . . [He would] give you a prescription to get rid of you. They really never did anything for me." As a teenager, her doctor gave her codeine, but it made her nauseous, so she made do with her mother's remedies instead. Later, she got prescriptions for Darvon and

Darvocet from her physician colleagues when she worked as a nurse. "That kind of helped . . . [but] I really hated to take any heavy sedatives because . . . they knock you out. So I was kind of cautious on what I was taking, depending on how bad I was." Once she convinced her physicians to take her seriously, they still were not much help. "They checked me out for all kinds of things. And they never found anything. So they always said, well, once you have a baby, it will probably ease up a little bit. But it really didn't too much after Carlen, but it did after the other two."

Rachel Cohen was equally unimpressed with her doctor's ability to diagnose her problem or do something useful about it.

> It was so horrible that my mother took me to a doctor. And there was nothing they could do. They prescribed Codeine for me. They actually prescribed Codeine for me . . . The doctor didn't ask me what I knew, what I didn't know. It was just a matter of prescribing something. Because he didn't know what was going on. Girls had pains with their period. I mean, that was pretty much standard. If you didn't, you were lucky, and if you did, you know, just take something for it.

Rachel was especially frustrated with her doctor's lack of initiative and knowledge in retrospect because she had concluded that she probably had endometriosis. Of course, endometriosis was not well understood, or particularly treatable, at the time, but Rachel was annoyed that her doctor did not even start with the "modern" belief that menstruation should be painless, and therefore did not try to figure out what was wrong.

Not every doctor gave prescriptions; Christina's just gave her advice. Calling on traditional thinking, he told Christina to back off of some of the athletics she enjoyed so much. "I do remember the doctor telling me not to play anything aggressively, to try to keep the body calm. Just go along naturally." She asked her doctor about discomforts other than cramps, as well. "My breasts would be sore as bananas, and I hated it. And then, I did ask my doctor about that, and I think he did say, full-breasted girls would feel it. And I did. You would feel it more." Some confirmation of normality might be reassuring, but it did not actually address the discomfort.

Mimi, Dorothy, Rachel, and Christina were probably lucky their doctors limited their advice and prescriptions. Some physicians publishing results and recommendations in early twentieth-century medical literature were confident that they could ease dysmenorrhea through operations to reshape and resituate the uterus, or through dosages of radium or electricity, though they admitted there were many difficult cases that did not respond to these treatments. Others tried various hormone supplements, without clear results. Nevertheless, they often displayed optimism about their ability to cure or reduce dysmenorrhea, and—along with physical educators, educational pamphlet writers, and industrial health workers—physicians acted as if a visit to the doctor would certainly solve any problem that regular exercise and good eating and sleeping habits could not resolve.[36]

As it turned out, doctors accidentally solved severe menstrual pain for some women by prescribing them birth control pills. Rachel Cohen started taking the Pill almost as soon as it came out, in 1960.

> I remember getting a prescription for Pills, and after that, boy my life was—I didn't even know to expect any relief from my period. I had no idea . . . all this torture I had endured for twenty years or whatever, fifteen years, I guess, was just suddenly evaporated. And I was just ecstatic. Because this changed my life.

Likely, her physician also had no idea it would help; it was simply a positive side effect. Mimi Cummings Wood was equally thrilled by the effect of the Pill when she began taking it after her marriage in 1968. "I never had cramps again . . . Had I known earlier that that would have resolved my problem! I just regret so much that I didn't, because my quality of life would have been so much better . . . It was miraculous. It was miraculous. I couldn't believe it." None of the doctors who examined Dorothy Joyce and wrote her prescription after prescription for sedatives ever suggested that she try the Pill to see if it would help. It is not surprising that physicians would have avoided recommending the Pill for managing menstrual cramps in unmarried women, but Dorothy was married for many of the years in which she was seeking help for her suffering. It does seem surprising that most physicians remained unaware for so long of a rather obvious side effect of the Pill, especially when the first FDA-approved use for the Pill was to regulate the menstrual cycle.[37]

The 1960s brought important ideological as well as technological changes. Over the course of the century, many sex education books expressed clear opinions on the danger or the value of sex during menstruation. At the beginning of the century, they warned about various risks to the health of female and male sexual partners. However, the advice changed markedly in the 1960s, when sexual mores changed rapidly, in the mainstream as well as in the counterculture.[38]

Beginning in the 1960s, most health and sex education literature advised that sex during menstruation was fine, as long as both partners were in agreement and there were no obvious underlying health problems that would be aggravated. A 1962 guide suggested that "having intercourse even during menstruation can be comforting and good," and a 1967 guide assured that "menstrual blood is perfectly harmless in content to both man and woman." It was suggested that sex during menstruation could even have beneficial side effects; a 1979 guide noted that "some women report the menstrual cramps subside after intercourse, particularly if the sexual experience has led to orgasm." These statements showed a radical shift from earlier medical and popular views and were part of the larger pattern of reduced health concerns surrounding menstruation.[39]

While these published commentaries were available to those who pursued them, most interviewees did not recall reading about sex during menstruation in books or hearing about it in sex education settings. Almost all women made a decision on their own about what they would do; few recalled discussing it with friends or with their partners. Several men born in the 1930s and 1940s recalled joking about it with male friends. They learned from their friends that it was not a good idea because it was unsanitary, or worse, if they risked their penises in a bloody vagina, "it's gonna fall off." Their friends warned that a woman might get angry if you tried to have sex with her during her period. These men had access to a particular discourse, which informed their thinking, even if it did not necessarily determine their practices, but the women in their lives mostly felt that they made decisions about sex during menstruation without assistance from books or friends. The men's talk remained among men and was not shared with their female partners. Most of those who did not normally have sex during menstruation had never actually had a discussion about it with their partners; both partners had assumed that it was not a good idea and avoided it except, perhaps, in exceptional circumstances.

Of women interviewees born between 1930 and 1960, most abstained from sex during menstruation, and some had not considered the possibility until it was mentioned in the interview. Unlike older interviewees, however, they explained that they abstained not because they thought it was unhealthy or "dirty" but because they thought it was "too messy." Abstention from sex during menstruation is not necessarily a matter of women's health concerns, their feelings about their bodies, or their fear of rejection by male partners; it can also be a matter of women's labor. Women have generally been the ones responsible for cleaning sheets, nightgowns, and anything else that might get stained as a result of sexual intercourse during menstruation. Just because women's health concerns surrounding menstruation had greatly diminished did not mean that they were willing to deal with the practical implications of having sex during menstruation.

A number of women who occasionally had sex during menstruation because of their partners' wishes explained that they tried to avoid it because of the extra mess, and therefore extra work, it created. Women had adopted all kinds of new technologies and practices in order to avoid manual labor associated with menstruation, so they naturally were not eager to embrace a new practice that involved that kind of labor. Amy Rivers, an African American and the daughter of a laborer and a homemaker, specified that her dislike for sex during menstruation was about messiness, not physical discomfort. "For me, it just wasn't a comfortable thing. I don't mean physically comfortable; it wasn't comfortable in terms of my hygiene. It would soil things; I didn't like that." Roberta Cummings Brown explained that she "concentrated more on 'what kind of mess am I making here?' " She detailed the work that followed a sexual encounter during menstruation. "We just take this linen off and put some more on and here I go. Rinse it all in the bathtub to make sure that I hadn't ruined my linen and nightgown or whatever." New Englander Margaret Olsen, daughter of a white electrical engineer and a homemaker, in addition to thinking, "this is really gross, watching his penis come out all covered in blood," found it annoying that her husband objected to her interrupting their sexual encounter to get towels to put down on the bed. The work in cleaning up after sex during menstruation could be significant, and many women found that it was not terribly enjoyable to have sex while they were worrying about making a mess and having to clean it up afterwards.

Male interviewees born between 1930 and 1960 mostly acquiesced to their wives' assessment that sex during menstruation was unacceptably messy or just generally went along with their wives' wishes. A few Chinese American women who had emigrated from Hong Kong explained that their husbands refused to have sex during menstruation because it was either "bad luck" or caused impotence. For the most part, however, it seemed that men of this age thought it was not inherently unreasonable to have sex during menstruation, and they did not mention the labor that accompanied the mess.

For some women, sex during menstruation was unthinkable because they were generally uncomfortable at that time, and sex simply was not appealing. Laura Hwang, an immigrant from Taiwan born in the mid-1940s, said that "you feel goosebumps so easily at that time . . . So I got trouble already, I don't want anybody to bother me." The partners of women who had this kind of discomfort took note of it. Laura said that her husband would not expect to have sex during menstruation because "he has to respect this kind of feeling." Discomfort could put sex off limits, even if concerns about messiness did not.

The sexual revolution and the women's health movement of the late 1960s and 1970s influenced thinking about sex during menstruation. Health and sex guides began to proclaim that there was absolutely nothing wrong with sex during menstruation, as long as both partners agreed to it. Both written materials and sexual practices generated during the sexual revolution affected how some interviewees thought about sex during menstruation. Laurie Wilson recalled reading in a book that the decision to have sex during menstruation was a personal choice and was perfectly acceptable, but found that "it was too far out of my mind to comprehend—it was still off limits for me." She never did adopt the practice, though reading about it made her realize that it was a possibility. Mimi Cummings Wood had also heard about the practice, and she explained, and almost apologized for, her choice in terms laid out during the sexual revolution of the late 1960s and the women's movement of the 1970s. She said that she abstained from sex during menstruation because she "never was that liberated." Peter Jefferson, an African American and son of a member of the Air Force and a teacher, attended college in the late 1970s. He described the woman who introduced him to the practice of sex during menstruation in terms that were familiar from the sexual revolution.

> I was dating a woman who was the last reincarnation of Earth Mother, and it was one day, we were making out or something or another, I don't really remember all the gory details, but it was like, "I'm having my period, but it's OK" . . . And so we went ahead and made love and it was like, wow! It was OK. OK! All those things those guys had told me about, you know, "It's gonna fall off."

As a result of the sexual revolution and women's movement, for some people, sex during menstruation became something done not by women who were forced into it by pushy husbands, as described by older generations, but by "liberated" women, some of whom adopted the "Earth Mother" persona, and their partners. Consistent with diminishing concerns about activities during menstruation affecting menstrual health and general health, interviewees born after 1940 either abstained from sex during menstruation because they thought it was too much bother to clean up, or they at least occasionally had sex during menstruation, especially if they allied themselves with the sexual revolution of the 1960s and 1970s. With regard to sex during menstruation, what the counterculture believed to be a rebellious act against repressive middle-class norms might be better interpreted as a sexual revolution or feminist inflected extension of those norms as they had been developing over the course of the century.

At the same time that sex advice manuals were beginning to advise that sex during menstruation was an aesthetic rather than a health or moral decision, in the late 1960s menstruation pamphlets and books began to tell pre-teen girls that they could truly do all activities during menstruation without worrying about their health. Experts no longer recommended against swimming or competitive sports or warned girls to avoid anything that would give them a chill. Among interviewees, most of those born after 1960 took this perspective for granted. They believed that they should be able to do all of their normal activities during menstruation, as long as they gave it some forethought and had planned appropriate management strategies. Chinese American Ginnie Wang, the daughter of teachers who immigrated from Taiwan, explained, "I only worry about it when I'm planning some trip and I have to think, OK, now I'll have to bring extra stuff with me." Carlen Joyce Thomas also would check a calendar if she was going to be traveling, to see "if I was going to be inconvenienced by it." She never planned a vacation around it, though. Neither did her sister, Serena Joyce Ambrose, although she preferred not to have her period in certain situations because of the extra annoyance it caused.

> We're leaving for camping on Friday. It doesn't matter if I have it, but it's a lot easier camping, to not. You're not running constantly to the bathroom. You know, you don't feel as clean all the time, and when you're camping you don't feel clean all the time as it is. This week I'm kind of looking forward to it, to get it over, so I can enjoy my weekend. Not that it's a big deal, but it would be nice to not have it, or have it almost at the end . . . I don't think I would change any plans.

Carlen described an incident in which menstruation had interfered in her athletic life.

> I was in a tennis match, and it was very close . . . I just happened to look down, and the inside of my legs were just, I mean, it wasn't dripping, but they were all red. I just completely lost my concentration, and I lost the match . . . it definitely distracted me . . . I was trying to fight, like, "I can

concentrate, I can stay in this match, I can win it, we're almost done, I can just win it and get out of here."

While Carlen had trouble finishing the match because of the visibility of her menstrual blood, she certainly was not worried that her health would suffer because she played tennis during menstruation. By the 1970s, women were being told that athletics during menstruation was safe, and they were also being told that they should be able to continue at full vigor all month. In the *Ladies' Home Journal* "Medicine Today" column in 1974, readers were informed that an American Medical Association survey of sixty-six female Olympic athletes showed that 75 percent trained as usual during menstruation, and only 5 percent stopped. This information was extrapolated to apply to non-Olympic athletes as well. "Physically fit women, the AMA notes, have less menstrual discomfort. So unless a girl's flow is excessive or her period unusually painful, the AMA says, 'there seems to be no medical reason not to train or compete during menstruation.'" In 1979, a columnist again talked about top-flight athletes, pointing out, "Medical surveys show that in Olympic competition 'women have set world records at all stages of their menstrual cycles.'" In the 1970s, the decade of Title IX's guarantee that women student athletes be given equal opportunities with their male peers, women's magazines ran articles encouraging young women to be athletic. As young feminists, they were encouraged to have faith in their own athletic abilities, to believe that women could be world-class athletes, like their male peers, and to see menstruation as in no way debilitating.[40]

To support all the activities they wanted to continue doing as usual during their periods, young women learned a great deal more from the mass media about potential menstrual discomforts and how to handle them than had previous generations. For young women in the 1970s through the 1990s, television commercials, Judy Blume, and educational pamphlets were only a portion of the media materials about menstruation they encountered. Women read about all kinds of things having to do with menstruation in magazines and widely distributed e-mail messages. Magazine articles were not new, but the range of topics surrounding menstruation that they covered was greatly expanded. Through the 1950s, magazine articles addressing menstruation were educational in the manner of school programs, focused on gynecological problems as health matters, or addressed sex education as a topic. By the 1980s, teen magazines began to publish articles about embarrassing moments and difficulties with tampons. Magazines aimed at teens and older women printed articles on managing PMS, exercises for cramps, the birth control pill, menstruation in relation to diet and general health, anorexia, Toxic Shock Syndrome, and menopause. Many of the articles were about managing menstrual discomfort that previously would have been regarded as minor or routine.

Interviewees remembered the topics of articles they read, even though they only occasionally remembered exactly what advice or new information they picked up from particular articles. Kim Chuang could recall stories in *Teen* and *Mademoiselle* about "how to feel better if you have PMS, abnormal bleeding, [and] funny stories about being on your period," as well as an article about a girl who died from Toxic Shock Syndrome. Felicity Chang never subscribed to teen or women's magazines or read them regularly; still, she mused, "I'm sure I picked up information, because you flip through, and there might be something on PMSing, or on complexion and hormone levels, so, it was definitely a source." Serena Joyce Ambrose subscribed to *Good Housekeeping* and *Mademoiselle* and kept an eye out for articles on PMS, from which she learned to avoid caffeine, salt, chocolate—"all the junk you want, because it only makes it worse."

All of this information was passed around among friends and was therefore spread much further than just to those who read it in magazines. Young women in the 1980s were more likely than their elders to talk to each other about what they had read or, by the 1990s, to pass it around by e-mail. Circulation of advice about how to manage menstruation was particularly common. Stacy Joyce Lockert's mother sent her magazine clippings and then they talked about them over the phone, eventually mutually diagnosing Stacy with PMS and figuring out together how she could get it treated. Heidi Xue never went looking for written information about menstruation, but

> I've been sent stuff. Actually, I think my friend sent me an e-mail recently. I think it's basically what you should eat when, that kind of thing . . . I think the main gist of it was if you experience cramping, or if you want to lose weight, there's a certain order that you should eat things, and I've never really felt I needed to change anything in terms of health or lifestyle, so I didn't retain much of the information that was imparted in that.

Another e-mail she received was a clipping "from one of the women's magazines, like Marie Claire, or Cosmo, or something, and it was all about being careful of who you date and when, because you're more susceptible to attracting not your type of guy at the wrong time of the month." She read the e-mails, even though she did not solicit them and was not inclined to take them seriously. The production and circulation of written stories and advice about menstruation was vast in the 1980s and 1990s, compared to earlier in the century, and gave women more ideas about what exactly required management and how they could manage all the different aspects of menstruation more carefully or effectively.

Young women who were interviewed largely embraced "modern" menstrual practices, regardless or ethnic or regional background. Young Chinese American women found, however, that unlike their peers with American-born parents, they were frequently in conflict with their mothers about their choices. Their mothers, immigrants from Taiwan, Hong Kong, and mainland China, had grown up believing that exercise and contact with cold water during menstruation could be dangerous. They encountered "modern" American menstrual management when they immigrated as young women, and while they enthusiastically adopted American ideas about menstrual education, they were less persuaded by late twentieth-century American ideas about menstruation and health. They were often baffled by their daughters' decision to exercise during their periods. Chinese American Angela Kong, the daughter of a physician and a homemaker, expressed skepticism: "One of my daughters say before her period comes, usually she really exercise herself, like a jog in morning. And she feels that helps her a lot with the cramps . . . I say, 'Oh.' I say 'I thought that's a time when you're not supposed to exercise.' [Laughter.] It's just opposite." Isabel Mao explained that in Taiwan, her physical education teachers had excused her from anything involving running and jumping, but

> I think nowadays, they think it's called old-fashioned because I ask my daughter, I said, she said, "No, we swim!" because she joined the swim team. I said, "You can't swim." I told her, I said, "When that come you cannot swim." She says, "No, no, no. We swim!" [Laughter.] I don't know how she does it. I guess now the teachers, they don't allow you, right?

She assumed that her daughter had made her own rebellious decision to exercise during menstruation; she could not believe a teacher would encourage that kind of behavior. Her daughter walked into the room during the interview, and Isabel explained that when she was growing up in Taiwan, during menstruation, "We can't run. We can't swim." Her daughter replied, "Yes. No, that's in the past. That's more from the Victorian era! [Laughter.]" Isabel exclaimed, "Oh, that's called Victorian. I see! Old-fashioned!" Her daughter explained further.

If I really needed to [get out of gym class], I could. But usually, unless you had real problems, you didn't need to. They encouraged you to participate as much as possible . . . I just used [menstruation as an excuse] once in my whole time in high school and middle school, and other than that, I didn't really see a need to, because I felt fine, too. I didn't have any problems.

Of course, for an American-born daughter who did not like gym class and had cramps during her period, her mother's health beliefs could be useful. Kim Chuang remembered that while her teacher tried to make her play in P.E. class all month, her mother told her she could sit out if she did not feel like it during her period.

In addition to telling their daughters not to exercise too much during menstruation, Chinese immigrants warned them not to drink or eat cold things and recommended they drink warm beverages. These recommendations came with varying degrees of explanation. Angela Kong learned from her own grandmother to avoid cold things but never asked why, so she could not offer a good explanation to her children. "You know, it's funny, I never find out. [Laughter.] They told me not to eat cold things, that's all. So I just stay away from them . . . Isn't that funny? Even I told my children the same thing. Of course they try not to, but I can see that they are very Americanized for the main, they'll just go ahead." Kim Chuang got the same advice from her mother and commented, "It's kind of cute. I'll probably do that to my daughter. 'You can't do that!' 'Why?' 'I don't know, because Grandma said it.'" Other American-born daughters were aware of their mothers' reasons for the advice but were not necessarily convinced. Amanda Chen explained, "My mom told me one time when I was in high school that I shouldn't drink really cold fluids or have ice cream or have something frozen when I have my period because it gives you more cramps because then the blood doesn't flow or whatever. I don't know, I pseudo paid attention to that. Not really." Brenda Xiao, daughter of an engineer and a homemaker, described, "She always made me drink hot liquids because I don't know if everyone thinks this, but the Chinese believe that when you drink hot liquids, it makes it flow through your body nicely or something like that." Despite this understanding of her mother's rationale, Brenda was glad to learn about Advil from her friends, and explained that, in contrast, "My mom kind of told me the wacky stuff, like don't use tampons or don't drink iced beverages."

With their limited understanding of Traditional Chinese Medicine (TCM), often gathered solely from parents who had left China at a relatively young age and had always mixed TCM and Western medicine, American-born daughters almost all assumed that they were simply supposed to avoid cold food and drink. They rarely considered the broader definition of "cold" foods within TCM, in which all foods, drinks, and medicines are "hot" or "cold" to some degree, some more than others. It is clear that at least some immigrants were thinking in these broader terms, but the meaning of "cold" was often lost in the translation to American culture. While their mothers were selective about which aspects of modern American menstrual management they adopted, the American-born generation rarely shared their mothers' qualms or even understood them.

———

Invested as they were in "modern" menstrual management, young women who were sometimes incapacitated during their periods could feel extremely bad about it. Maggie Yi, the daughter of physicians who opened a restaurant when they arrived in the United States, struggled with painful menstrual cramps throughout high school and her professional training. She described both why it was so important for women to deny that menstruation could be problematic and why it was profoundly disturbing to her that she was sometimes debilitated during her periods.

> I remember actually reading some article which was highly offensive, the kind of thing where you, part of your brain immediately dismisses it, and part of you is really bugged by it. And just reading all this stuff about how . . . women wouldn't be good at various things because even if they were the most brilliant person on earth and everyone agreed that they were so amazing, they were normally very capable, there would be these, potentially these blank periods, and what if that was the time when they were deciding to launch nuclear weapons, or something. It was basically along those lines. And I remember being like, "That is the stupidest thing I ever heard," but part of me wondering if I'd be able to handle really, really high-pressure situations. If there just came a day where I really felt like I couldn't exist at all, anywhere. What would I do? . . . I was sort of like, "That's just so untrue. I don't believe that at all." But . . . [in a way] I did.

The "modern" mode of menstrual management developed by their foremothers was of tremendous importance to women born during and after the feminist movement of the 1960s and 1970s, who were raised to believe they could take on any career but nevertheless occasionally faced public suggestions that menstruation would prevent women from attaining the highest ranks of business or public life. Maggie Yi found herself caught in the middle: she could not manage menstruation perfectly, let down as she was by the technologies and experts of menstrual management, but she was angry at the suggestion that menstruation should disqualify women from positions of responsibility.

Young women like Maggie, who did everything they could to manage menstruation so that it did not impinge on their lives, could be quite critical of "experts" who did not do their expected part in supporting modern menstrual management. Women of this generation were willing to criticize male and female doctors alike, when they felt they mishandled discussions of menstruation, especially as it related to reproductive health. They had much higher expectations of their doctors in terms of discussing menstruation and treating menstrual problems than had previous generations. Amanda Chen expected to discuss menstruation with her doctors in the context of regular checkups. "They usually ask, "Are you regular? Are you having a problem with your period?" She was critical of a physician who did not initiate this kind of discussion. "I just had [a gynecological exam] in December, and it was the first time I had been to this doctor, and she was late and all that good stuff. She was very brief and didn't ask the usual questions and stuff like that. Which I was not too happy about." Lisa Wood, growing up in rural Virginia, never discussed menstruation with her doctor in a way that she felt was adequate.

> My family doctor since I was born is a male. I don't remember talking to him. I would go to him at least three times a year. I don't remember him ever asking me about my period. Maybe, "Is it OK?" . . . What's the definition of OK? . . . It could have been completely off kilter, for all I knew. Because I didn't know what "OK" was.

Women born after 1960 not only were critical of doctors who failed to understand thoroughly and address normal menstruation and its relationship to reproduction but also resented doctors who failed to take seriously self-diagnosed problems with menstruation. Stacy Joyce Lockert, dealing with extreme mood shifts over the course of each month, went to her gynecologist to discuss them and spent months persuading him that she suffered from a condition related to menstruation that should be regarded as medically serious.

> It took me a while to get him to take this seriously. And actually, I had to go to a therapist to figure out what was wrong, and that's what they came up with, [a diagnosis of PMS] . . . I had to keep track, in a notebook, of my

moods, when it started, the two weeks of my PMS, then actually for the whole month. My moods were completely different—we're talking black and white. That wasn't me, it wasn't normal, and so I had to address my doctor after I addressed my therapist, with the kind of test we did, that we had to take this seriously. The gynecologist didn't prescribe anything for me; I ended up getting something through the psychiatrist.

Not all women who diagnosed themselves with some level of PMS were dismissed by their doctors. Serena Joyce Ambrose discussed it with her doctor every year and accepted the medical advice to watch her diet carefully and exercise, even though she felt this recommendation was not a spectacular solution but rather a way to "just kind of get through it." She was not looking for a prescription, but she expected her doctor to provide some advice and ideas about how to cope with her problem, and her doctor was obliging. While many doctors in the 1980s and 1990s were perfectly willing to discuss menstruation with their patients, and offer ways to help them cope with menstrual pain and PMS, as well as counseling about what constituted "normal" menstruation, they do not seem to have been eager to medicalize menstruation. If anything, these women's doctors seemed less inclined than the women themselves to regard menstruation as something which should be managed medically, which perhaps is unsurprising given the continued lack of solutions the medical profession could offer for menstrual pain that did not respond to self-treatment, aside from birth control pills. While a few doctors made splashy, well-publicized cases for the medicalization of menstruation, especially in relation to PMS, they do not appear to have represented the profession as a whole.[41]

———

Interviewees born in the last decades of the twentieth century expected support, assistance, and sympathy not only from doctors but from family members, friends, and even colleagues. These young women expanded significantly a pattern instantiated by their mothers, looking to husbands for sympathy as well as to men outside their immediate families. They were willing to manage their menstruation assiduously, but they were not as willing to hide their management efforts as their mothers or the educational pamphlets of the 1940s and 1950s might have advocated.

Young women struggled with, and were angry about, any suggestion that menstruation should keep them from full equality with men, but at the same time they were dismayed at the thought that the annoyances and discomfort of menstruation should be hidden and therefore go unrecognized. Inheritors of the school and workplace breakthroughs of their feminist foremothers, these women were generally confident of their place beside men in their studies and their professional ambitions, and they did not think that they needed to deny all biological difference from men to compete effectively. They seemed to believe that as long as they were essentially as productive as men, they should not need to pretend to have bodies that were exactly like men's, and their attitude toward the modern period reflected this conviction.

Young women felt they had support from their peers in taking this stance. As lawyer Ginnie Wang put it,

> Seeing more people around me talking more comfortably about the issue, and kind of feeling, well, you know, this is not something to be embarrassed of, or to hide, because it happens to half the population in the world, and it's part of what makes us a woman, and if other people can't deal with it, tough. And after that realization, I think it got a lot easier to accept it as something that just happens. If it happens, and I feel crappy that day, I will tell someone and tell them to just kind of leave me alone. As opposed to trying to cover it up or hide it or trying to pretend that I'm not feeling lousy. So I think it's more of an acceptance, an attitude change, that's made it a lot easier to deal with.

Many in this generation started by talking about menstruation at home, first to their mothers, then to their male relatives. African American John Graves remembered that his daughter "always felt free to talk, I guess . . . She's gonna say, 'I have cramps today.' Rarely would she stay home from work. But she would say, 'I thought I had better stay home, I think I'm catching a cold, and I have cramps, too.'" She continued to talk to him about it freely as an adult. "She still lives here, and works here. And she'll call you now! You know, if something is going on, she'll let you know."

Young women were also much more likely than their mothers had been to mention menstruation to their brothers. Keith Zhao recalled of his sister, "Every so often she would come up and say to me, 'you know, I'm having my period, I'm going to go lie down,' or something like that." It was not necessarily at moments when she felt pressure to explain why she could not do something; she just wanted to complain to her brother. "She would just mention it off-handedly." Justin Tsai, the son of accountants, remembered menstruation coming up in his telephone conversations with his younger sister when she was at college in the late 1990s. She and her roommates had synchronized cycles, and she would complain to him, "Oh, God, I can't stand it, next week is going to be period week."

This pattern of complaining to men about menstruation or asking for assistance in managing it extended outside the home, to male friends, boyfriends, and colleagues. While young men did not always feel comfortable talking about menstruation, young women insisted on telling them about it, and men responded in a variety of ways. Traci Anderson explained the reaction of her college boyfriend in the 1990s.

> I think it used to freak him out a little bit, as it does a lot of guys. It would mostly come up because he would come over, and I would be lying in a miserable ball on the bed. He'd say, "How are you?" and I'd say, "Cramps!" "Oh, I'm sorry." He'd be sympathetic, but he wouldn't really want to go into it. I think that's probably fairly typical.

Traci was willing to let her boyfriend avoid talking about it, but others were not so resigned. A number of sexually active young women expected their boyfriends to be comfortable having sex during menstruation. The changing practices of the 1960s and 1970s, and perhaps the changing advice that reached more people through school education programs, seems to have had an influence on many young women's expectations. Lisa Wood said it had not occurred to her when she first became sexually active that some people might object to sex during menstruation. "The first people that I was with were just completely and totally open and accepting ... I'm sort of annoyed by people who have problems with it, or problems with body issues in general." She thought about it in terms drawn from the 1970s women's movement, considering abstention from sex during menstruation to be a rejection of women's bodies. Not all young men were comfortable with the idea of having sex during menstruation, though. Among male interviewees, Chinese Americans were particularly unlikely to be willing to have sex during menstruation. It created awkward situations when their female partners suggested it, and they either went along reluctantly or outright refused. Women who had decided they could deal with messiness in order to be liberated found themselves feeling annoyed and rejected, and men could not figure out a graceful way to approach the issue when it was clear that they were not going to be considered chivalrous for abstaining from sex during menstruation. For some couples, sex during menstruation became a point of contention in a new way.

Young men's reluctance to support young women's "modern" management of their menstruating bodies was apparent outside of intimate relationships as well. This reluctance did not, however, keep young women from demanding recognition of their management efforts from male friends. Like older women in her rural African American community, Anika Taylor, the daughter of a truck driver and a homemaker, faced teasing from young men about menstruation, but unlike Roberta Cummings Brown and other older women, she responded assertively.

> My male friends don't want to hear anything about it. They're so sad. It's like it's the worst thing to hear. They're like, "Oh, that's so nasty." I'm like, "Shut up!" They make it sound so bad. Because I do have some close male friends . . . They don't want to be bothered . . . But I say, "You're old enough. You're acting like you're twelve, or something." They're like, "OK, that's enough." But I don't have no problem talking about it.

Alex Jones experienced this new assertiveness from the receiving end. A school sex education class about menstruation in the early 1990s led to a group effort by girls in his class to educate boys about the experience of menstruation. One girl, with the support of her friends, told the boys, " 'You don't know how bad it hurts. Imagine if somebody kicked you in between your legs, and you walked around the whole day!' We were all sitting around, like, 'Man, I don't know what you're talking about.' " He thought that something similar had happened in his high school sex education class, in which the discussion "went outside the textbook, and girls were being kind of personal about it . . . 'You all don't have to worry about carrying around all these things in your purse!' " The teacher laughed, but let them have the discussion. "I think it actually made people start talking, and we got open about it." In his high school, girls also challenged the boys' teasing in less formal settings. "By senior year, girls were more open about it, saying things like, 'Oh, you boys have the same thing, you all just don't realize it. You're moody at times too.' " There would be public lunchroom conversation or debates, "little wars" between girls and boys about it every once in a while. The boys were perfectly comfortable teasing the girls, but when the girls insisted on answering back, the boys would find ways to shut down the conversation: "Somebody would be like, 'Ew, that's nasty, man, you all stop talking about that at lunch.' " This type of conversation continued after college, at work, for some. Kim Chuang, working in a high-tech startup in the late 1990s, was open with her young male coworkers about menstruation, telling them when she had cramps, even though "They're like, 'We don't want to know this.' "

Unlike the previous generation, men were more often included in serious conversations, not just when they were teasing or joking. Alex Jones explained that in college, "It's not a laughing matter anymore, like it once was, and girls feel it's understandable, between everybody, now. About that age, they probably figure, if you don't understand it, you're probably really immature." The kind of education his female peers in high school had insisted on giving the boys had sunk in by college, and college women could mention menstruation without making it a battle or a lesson. "Sometimes they'll be explicit—'Oh, man, I've got cramps,' or something. They'll just say it . . . You'll see them grab something, a pill, out of their purse. Or they'll be like, 'I'm going to my dorm room, I'm going to sleep' . . . It's not an everyday thing, but you do hear girls say that every once in a while." Adam Chiang, the son of an engineer and a homemaker, found that at his northern California university in the early 1990s, discussion about menstruation "was always in the context of PMS . . . It's just part of people saying, 'So-and-so is like this because she's—[got PMS].' Just because other women are aware of other people's cycles, so they'll discuss it. That's the first time it was an open topic." The newly popular subject of PMS was an integral part of the increased volume of discussions between women and men.

A British physician, Katharina Dalton, and a fellow researcher created the acronym PMS in 1953. In creating the term, Dalton was modifying and expanding the category of "pre-menstrual tension," proposed by physician Robert Frank in 1931. Dalton published a book, aimed at a popular audience and titled *The Pre-Menstrual Syndrome* in 1964, but the term *PMS*, and attention to it as a phenomenon, did not really catch on in the United States until the 1980s. Feminist sociologists have attributed the rise of interest in PMS to two sources: first, two controversial and well-publicized British murder trials in the early 1980s, in which the female defendants' sentences were mitigated because they were found to have been suffering from PMS at the time they committed the murders; and second, a generalized cultural and political backlash against feminism, part of which came in the form of arguments that women's biological limitations would inevitably undermine their attempts to undertake high-level professional work and successfully serve in positions of power.[42]

These factors undoubtedly contributed to the rising popular interest in PMS in the 1980s. However, the emergence of PMS as a focus of popular concern should also be placed in the context of the longer twentieth-century history of the management of menstruation as a means of envisioning and realizing "modern" bodies. Contextualizing PMS in this broader history of bodily management not only provides an additional, historically sensitive explanation for the emergence of PMS as a popular topic in the 1980s, it helps make sense of the (perhaps surprising) ways young women and men would come to talk about PMS in the 1990s.

Women's thoughts about PMS reflected the beliefs and practices surrounding menstruation of the time in which they grew up. Among the oldest generation of women and men interviewed, mood changes were associated with menstruation only insofar as being "on the rag" could make someone uncomfortable enough, or cause enough physical pain, to put her in a bad mood. Some of the oldest women interviewed did not know what PMS was when they were asked about it. Many, especially Chinese Americans, had heard of it, but felt unconfident talking about it, asking for a clarification of the term first. Those who had some knowledge of PMS thought of it as a reference to a change in mood and were generally very skeptical about it. Mary Hanson was particularly harsh.

> I think that's a big joke. Well, you know, I suppose that reverts back to when you didn't talk about those things to your mother. You didn't talk about your period or what happened. Now it's, "Oh, it's because" you "had your period." You never heard that expression. So you didn't associate it with anything at all. If you had a headache, it wasn't because you had your period. And if you were cranky, that wasn't attributed to that. It was your bad disposition. Which is true! I mean, they blame a lot on that, which I think is a crock. They use it.

Mara Ozols was less critical of those who might talk about PMS, except in the context of television advertising, but she remembered her surprise when she found out what it was.

> I stop [menstruating] already when I start hearing PMS. What PMS? [Laughter.] Now I know, but I never heard PMS . . . My gosh, they advertise too, PMS. Medicine for PMS . . . first was completely, completely blank. Then later I start reading, start see magazines. Oh, that's what PMS! I start thinking, I say, "If my mood change, I don't know. Nobody told me!"

Interestingly, Mary and Mara point out that they did not associate mood changes with menstruation, not just because they did not remember experiencing them but because no one told them that they should expect to experience them in a particular relationship to menstruation.

New modes of education, new ideas about menstrual fitness and health, and new menstrual technology, initiated in the early decades of the twentieth century and elaborated over the rest of the century, would provide the grounds for PMS to become imaginable and recognizable. Women became convinced that they ought to be able to get menstruation under control if they only employed the right technologies, techniques, and attitudes. They employed increasingly sophisticated modes of menstrual management accompanied by increasingly stringent expectations for themselves and others. This trajectory would culminate in a late twentieth-century discourse about PMS that would flip-flop between refusing to strive for perfection in menstrual management and, on the other side, insisting that it be extended far beyond what could have been imagined at the beginning of the century.

By the time women born in 1950 came of age, "modern" menstrual management was well established in American culture. Most girls learned about menstruation before menarche and encountered educational programs and pamphlets within a year or two after that if not before; tampon use was common, even for unmarried women; and the Pill was a new and increasingly popular birth control option with the convenient side-effects of making menstruation relatively light and extremely predictable, and sometimes relieving menstrual cramps. In 1984, in the middle of these women's menstrual lives, Ibuprofen was approved for over-the-counter use and was widely adopted for quelling menstrual cramps. All of this technology was relatively inexpensive and accessible, and most of it was used by the majority of American women. Just as most Americans had come to consider themselves part of the middle class, most American women had largely attained a "modern" menstruating body, free of leakage, odor, and inefficiency, which they felt supported a middle-class lifestyle and self-presentation.

Once all these modes of menstrual management were in place, women were ready to take seriously the proposal of yet another way they should better manage their menstruating bodies, keeping emotions as well as blood flow in check. The "discovery" of PMS was not the first time that bad moods had been linked to menstruation: nineteenth-century medical writers believed that all female sexual and reproductive functions, including the menstrual cycle, had a strong impact on the nervous system. Consequently, they regularly associated the premenstrual and menstrual periods with emotional instability and irritability. In the twentieth century, medical writers concerned about cycle-related emotionality attributed it to hormones.[43] Among interviewees, and many other nonmedical twentieth-century sources, however, it was largely assumed that women were grumpy and irritable because menstrual pads chafed and menstrual cramps were painful, not because emotionality was inherently linked to the menstrual cycle. But by the early 1980s, women were more inclined to take seriously the suggestion that moods could be associated with the cycle itself, given that the physical discomfort that they assumed caused bad moods during menstruation had largely been eliminated. They were ready to be convinced that periodic irritability and depression should be seen as a part of the menstrual cycle that needed to be brought under control, and they read with interest the abundant popular literature about premenstrual syndrome.

Both women's magazines and general interest publications presented articles detailing the symptoms attributed to PMS, often including quotes straight from the syndrome's biggest promoter, Dr. Katharina Dalton, about the terrible tolls PMS took on women's family life and work productivity. Following Dalton, popular articles mentioned a long list of physical symptoms associated with PMS but emphasized psychological symptoms such as depression and irritability as those that most disrupted women's productive and reproductive efficiency and therefore most needed to be managed. Estimates of the number of women who suffered from PMS varied tremendously, depending on which authority was being cited; many articles claimed that the majority of women suffer from at least some degree of PMS and could benefit from advice about diet and exercise, and possibly prescription medications.[44]

Most of the women interviewed who were born in the 1940s and 1950s, and who had so assiduously adopted "modern" menstrual technologies and practices, were surprised when they began to read about PMS in the 1980s but generally were not skeptical about it. Anabelle Qian, born in the late 1940s in Taiwan to an air force general and an accountant, was initially surprised but was basically willing to apply the concept to her own experiences.

> I was just thinking, well, some people have it, and some people don't, right? So I guess I thought I was one of the lucky ones. I don't have too serious problems, but I think some people may have stronger symptoms. I think it's real. Maybe it's something like my cramps; it's just manifested in different ways. But personally I don't have any experience. Well, I don't know, sometimes, if you make too much out of it . . . but I think it's probably a positive thing to identify it, because now, come to think about it, sometimes . . . If I do feel down, and then it was getting to my period, I just kind of attribute that to the PMS. I don't know if it's just coincidence, or caused by some physical uncomfort. When you physically don't feel all together, then psychologically you can feel kind of depressed, or just feel a little bit down about things.

Anabelle and others agreed with Barbara Ricci about what the appropriate response should be to the mood changes they attributed to PMS.

I think there probably really is something to the fact that there's mood changes, and it's physiological. But I think we should try to carry on a normal life anyway, in spite of it, rather than using it as some kind of an excuse for murdering someone that day. [Laughter.] I think it's good, also, that we recognize such a thing. For sure. And I think hormones do make a difference in how we think, in our moods. So I'm glad everyone knows about it and can accept it . . . but I don't think you should plan your life around it. Because then I think you're thinking too much about it, and that wouldn't be good. So, I'm glad we're so understanding about PMS, and know what it is, but I don't think we should use it as an excuse for bad behavior or changing our lifestyle.

Roberta Cummings Brown, looking back on her daughter's painful experiences with menstruation, interpreted them as PMS. This interpretation inspired in Roberta more understanding, though not entirely more sympathy with her daughter's behavior. "She got deathly sick. She threw up, she had cramps, and she was out of commission every day of her life until she had a baby. One day out of every month. She was really hateful. She had PMS in the worst, the worst case of it . . . I didn't know the name of it, but I knew it was something bad." When asked what she had called it at the time, Roberta exclaimed, "A bad attitude. [Laughter.] It was like, 'You just have a bad attitude! You just have a bad attitude, and there's nothing I can do about it!' . . . I felt bad, but she's just impossible to live with." She was not so certain that naming and medicalizing this behavior as PMS solved the problem. "I understood psychologically what was really going on. I think I did anyway, but I thought that she was supposed to have more coping skills."

Those who felt they really did suffer from PMS also agreed with Barbara that PMS was no excuse for bad behavior or alterations in normal activities and sought advice from their doctors. They worked to implement their doctors' recommendations to alter their diets and exercise patterns, accepting that cyclical mood changes were a part of the menstrual cycle that ought to be carefully managed.

Men of this generation were less sympathetic to the idea that PMS symptoms reflected physiological changes associated with the menstrual cycle. Peggy Woo had debated the point with her husband. "My husband asked me if it's a real thing. He doesn't believe it is. He thinks that it is a big excuse for people's actions, and they way they act and react, another thing to sort of, worry about, so it's an intangible, to him, disease, or state." Joseph Wong, born to an architect and a chemist in the early 1950s, was finally convinced of its existence by his partners.

"I think for the longest time I didn't think that existed. I think it definitely exists now. All these partners can't be wrong. [Laughter.] Some have it more severely, others don't. I think it definitely exists, and for me that's a major change of mind."

While Joseph and Peggy's husband had to be convinced of the existence of PMS, other men were perfectly willing to treat it as something real if they wanted to criticize their wives or coworkers. Susan Ozols and her coworkers dismissed their husbands' comments. "Some of the women would say, 'Oh, God, my husband says I have my PMS now, because I'm grouchy, and it's around my period.' Stuff like that. Then we just kind of laugh, and figure, 'Oh, well.' [Laughter.] 'They'll survive.'" Susan could laugh it off, but the women who worked with some of John Graves's colleagues may not have been so sanguine. "There were some people who didn't believe that females should be in management because of that. That was one of the issues . . . You might hear some real crude-ball talk about it, maybe trying to talk about a business problem, but he would bring that in." Men of this generation held women responsible for managing their mood changes, either refusing to believe they were associated with the menstrual cycle, or blaming mood changes on the menstrual cycle and insisting that women control them if they wanted respect at home or at work.

Women born in the last decades of the century, however, initiated a new pattern in discussing PMS, more generally in line with their management of menstruation. Once *PMS* had been made available as a term, young women and men modified it to perform a variety of functions, some only tenuously related to premenstrual syndrome. The term *PMS*, as used by young men and women in the 1990s, had two main, and opposing, functions. First, it provided a way for young women to insist that their managerial work around menstruation be recognized. Second, it provided a way for women and men to insist that both women and men's unreasonable bad moods be subject to management.

To fulfill these functions, young women and men took this medical term and turned it into a verb, *PMSing*, which African American Ross Lyons, son of a self-employed businessman and a banker, referred to as a "slang term." With its medical lineage, it became an accepted way for young women to mention menstruation, something previously unmentionable, in relatively public settings. Some young women used it in ways that referred directly to 1980s popular understandings of PMS. Felicity Chang observed, "I think the first time I started talking with anybody about [menstruation] was maybe in college, and it was because at some point it started to become very common for people to talk about

PMSing, and it was just a very open thing, like, 'Oh, whatever, I'm bitchy because I'm PMSing.'"

For this generation, however, the understanding of PMS was a lot broader than either the medical definition or their parents' use of the term. As Keith Zhao explained,

> I've always equated the words *PMS* and *period*, and kind of all those terms. And so, I guess I've never really differentiated, like, cramps that my girlfriend may have had during or after menstruation, from PMS, and your questions up until now kind of suggest that there's a different kind of thing that happens before menstruation, that there may be controversy over. And I just realized that. So, just to let you know . . . My sister has also said, "I am on the rag," and "I am having PMS right now." So taking all of her statements together, I kind of lumped them all into the same term.

Keith's broad understanding of the term *PMS*, encompassing everything about menstruation, is not an example of male ignorance or uncertainty. Rather, his understanding reflects how both women and men of his generation used the term, especially when talking to each other about menstruation.

Anika Taylor used the term when she had cramps during her period or was feeling reserved about physical activity because she was not using tampons, to explain to her male friends why she was quieter or less enthusiastic about their activities than usual.

> If I have to go do something with [my male friends, they'll say], "Well, hopefully you're not PMSing, because I don't want to . . .' I am like, "Oh, my goodness, you are so wrong." They don't want to hear nothing about it. They think it's the worst thing to talk about. They're like little boys. They're like, "Ew, don't tell me nothing about it, I don't want to hear.". . . I'll be like, "You asked what's wrong with me." I'll tell them, and they'll be like, "All righty, thanks for telling me your business." I'm like, "You asked!" I figure, they're old enough. They're twenty, twenty-one; they should know what's happening. Especially if they have a girlfriend.

Talking about PMSing was a way that Anika could insist that male friends hear about the discomforts and the management efforts surrounding menstruation.

On the flip side, when the term *PMSing* was used to talk about other people, it generally turned into a way to insist that women manage their moods better rather than a way for a woman to demand recognition of her management efforts.

Traci Anderson worried that "men can use it negatively against women to just say, 'Oh, it's just PMS', like 'Oh, she's being a bitch' or something like that." Some young men did indeed find it to be a useful slang term, in just the way Traci described, because while it was often used as a substitute for "on the rag," as in, "she has an attitude because she's PMSing," the term seemed to escape the lower-class connotations of "on the rag."

As it came to be used, however, it was not simply a middle-class way for men to derogatorily attribute women's anger to their biology. It also somehow escaped the gendered boundaries implicit in the term "on the rag," and women and men alike used it to insist that men's anger and moodiness also be well managed. Ross Lyons explained,

> That was a slang term that we probably picked up before the [sex education] class. You know, just, "He's PMSing." Nobody really knew what it meant until later on in life . . . it was more of a, "Oh, he/she's acting up." And [we] even used [it] for guys, so it wasn't a sexist type thing . . . They just might say, "Oh, he's PMSing." He's like complaining, crying, he's upset all the time. Just leave him alone and let him be. That's more what it was used for. "She's PMSing," is just like leave her alone, leave him alone, whoever.

Of course, the implications are different in using the term to describe women and men. Saying that a man is PMSing is a way of teasing him out of a bad mood by threatening his masculinity, while saying that a woman is PMSing is a way of dismissing her bad mood as determined by her biology. Still, it is intriguing that Ross Lyons reported learning the term on the playground as a child, from other boys, before he had any inkling of its gendered implications or its link to women's bodies. A mode of managing women's "modern" bodies may be expanding to include men's bodies as well.

PMS holds an interesting place in the history of the development of the modern menstruating body. As a popularized medical concept, it was largely accepted by women in the 1980s as yet another mode of menstrual knowledge and management. In the 1990s, however, as it crossed the boundaries between the medical and the popular, the unmentionable and the mentionable, and female and male, it became useful both for young women insisting on recognition for menstrual discomfort and the annoyance of management practices around menstruation, and for young men and women expanding the realm of menstrual management to include male bodies.

In the century between Jacobi and Clarke's debates about whether women would ruin their health by attending school during their menstrual periods, and young interviewees' use of the term *PMSing* to insist on thorough menstrual management and the recognition of management efforts, the modern period emerged. Physical educators, industrial hygienists, and educational pamphlet writers encouraged women—sometimes with a strong element of coercion—to treat the days of their menstrual periods just like any other time of the month. They and the women they counseled abandoned long-standing health concerns about exposure to cold and emotional shock during menstruation, focusing on how to avoid menstrual discomforts that slowed them down rather than on precautions to protect their health at a vulnerable time. Experts began to advise that sex during menstruation was safe and reasonable, and while interviewees born between 1930 and 1960 generally abstained, they did so to avoid the labor of cleaning up the mess rather than out of concerns about health or morality.

Women sometimes went to doctors for help with menstrual pain, although they were unlikely to receive much effective help until the advent of oral contraceptive pills in 1960. When they did have pain or discomfort they could not relieve, women born in the 1940s and 1950s, the most enthusiastic adopters of modern menstrual management, pushed their definition of appropriate management in a slightly different direction than the "experts." As far as these women were concerned, being modern did not mean just *appearing* to not be menstruating and hiding discomfort that could not be relieved; rather, they began to expect sympathy, even assistance, from their husbands in managing discomforts they could not erase on their own.

The youngest women interviewed for this book, born during and after the sexual and feminist revolutions of the 1960s and 1970s, took for granted the modern perspective on menstruation and health. This was true even of the daughters of Chinese immigrants, who rarely heeded or even really understood their mothers' concerns. They read a wide variety of advice about how to manage many aspects of menstrual changes and were highly critical of doctors and other experts who did not support their efforts as well as these young women believed they should. They also expanded the range of people from whom they expected sympathy and support to include many more men who were not family members.

New beliefs about health and menstruation and new ways of educating girls and women about menstruation came together to created "modern" bodily management. Modernization in these two realms, however, was not the entire story. It would also take the widespread manufacture, advertising, and adoption of new menstrual technology to make it feasible for women to put into practice their new beliefs and expectations about menstruation.

CHAPTER FOUR

The Modern Way to Manage Menstruation
Technology and Bodily Practices

New technology introduced on the emerging mass market in the 1920s and 1930s, and advertised aggressively in nationally available media, transformed women's and men's experiences of menstruation. Advertisers presented a vision of a "modern" body that was well managed, did not leak or smell, did not cause anxiety or self-consciousness, and did not display other evidence of menstruation. They intimated that realizing this vision was necessary to attaining middle-class status. As a 1937 advertisement put it, "At first you will not be able to *believe* in the freedom . . . the comfort . . . the poise Tampax makes possible. And soon, like thousands of others, you will wonder how you ever existed before this civilized method of sanitary protection was perfected."[1] Women eagerly adopted disposable sanitary napkins and somewhat more cautiously experimented with tampons. They found that these technologies addressed long-held concerns about comfort and reliability and at the same time supported the "modern" self-presentation so eloquently articulated in the ads. In order to take advantage of the potential increased convenience and discretion, women, advertisers, retailers, and others had to create new public venues where menstruation could be acknowledged. Just as classrooms and libraries had to become places where menstruation was discussed in order to adopt a "modern" scientific explanation of menstruation, magazines and newspapers, drugstores, bathroom cabinets, and trash cans had to become places where menstrual products could be seen in order for women to adopt "modern" technologies. Modern discretion was paradoxically based on greater public acknowledgement and display, albeit in carefully circumscribed locations. By the end of the twentieth century, young women happily took mass-market technologies for granted and began to expect boyfriends, relatives, and even colleagues to tolerate public display of menstrual technology outside the bounds established by their mothers and grandmothers and to even occasionally help procure menstrual supplies.

Advertisements for Kotex and other sanitary napkins, introduced in 1921, and tampon advertisements, beginning in 1936, articulated clearly, and publicly, a vision of the "modern" menstruating body. This vision was developed in the ads over the course of the 1920s and by the 1930s had taken on a form that would remain familiar through the end of the century.

As T. J. Jackson Lears has shown, ads for products that assisted in bodily management, such as soap and deodorant, began to proliferate in the late nineteenth and early twentieth centuries, and changed character significantly in the first three decades of the twentieth century. Nineteenth-century ads for soap showing scantily clad women in exotic locales gave way to "standardized, sanitized images of youthful physical perfection" in more everyday settings. Cures for constipation and other bodily discomforts and inefficiencies were no longer promoted with references to mystical formulas obtained in far-off lands but instead on the basis of scientific and medical authority. The language of efficiency and productivity, adopted so eagerly in industry during this period, was taken up by advertisers to urge consumers to manage their bodies as carefully as industrialists managed their factories.[2]

This change in the character of advertisements for body management products signaled the beginning of advertisers' self-conscious attempts to shape Americans' images of themselves. Advertisers began to link images of recognizably middle- and upper-class women and men, participating in clearly middle- and upper-class activities, to text that used the language of "modern" science, medicine, and business to argue that their products were necessary to successfully perform these activities and consolidate modern versions of these identities. Manufacturers of menstrual technology were enthusiastic contributors to this new kind of advertising. Early ads for Kotex, beginning in 1921, were aimed at an upper- and middle-class audience who already had embraced the regular use of soap and baths in order to be "decent," and had learned to worry about germs and to value "hygienic" and "sanitary" items and practices.[3] At the same time, these ads consolidated these values and practices into an emerging vision of a modern, middle-class female body, juxtaposing images of glamorous women in evening dress with inset images of efficient typists, and copy that emphasized the convenience and health benefits of a disposable product.[4] As the established middle class redefined itself around claims to professional and managerial expertise, and the aspiring middle class flooded into the growing number of newly created clerical and sales jobs, manufacturers and advertisers were determined to make new menstrual products indispensable in attaining the self-presentation these reshaped identities required.[5]

Other Women Will Tell You

how this new form of sanitary protection brings freedom, comfort, better health

A scientific method replaces hazardous, dangerously uncertain makeshifts—and brings new standards of health, comfort, fine grooming.

BECAUSE science has found a solution to woman's oldest hygienic problem: because doctors and nurses have discovered this new form of sanitary protection and women have told one another of its remarkable advantages—the hygienic habits of women have changed all over the world. Today, one takes quite for granted the comforts of Kotex, which plays so vital a part in new-found feminine freedom and peace-of-mind.

Mental as well as physical relief

Within the past year, further refinements have been made that assure greater mental as well as physical comfort.

Kotex has always offered these advantages:

Cellucotton absorbent wadding, the super-absorbent filler, takes up 16 times its own weight in moisture; it is disposable, just like tissue ... easily, quickly; wrapped in soft, specially treated gauze, it offers gentle, easeful protection; the filler is adjustable — layers can be removed to suit one's individual needs.

Use Super-size Kotex
Formerly 90c—Now 65c

At the new low price, you can easily afford to buy Super Size. Disposable in the same way. Super Size offers the many advantages of the Kotex you always use, PLUS THE GREATER PROTECTION which comes with extra layers of Cellucotton absorbent wadding. Doctors and nurses consider it quite indispensable when extra protection is essential. Buy one box of Super Size to every three boxes of regular size Kotex. Its added layers of filler mean added comfort.

Deodorization and a new cut

In Kotex Laboratories a way has finally been found to deodorize Kotex safely, completely. This process has been patented and is to be found in no other sanitary pad. Then, too, corners have been rounded and tapered so as to leave no evidence of sanitary protection.

A recently developed treatment of both gauze and filler adds to their softness. If you have not yet used the new Kotex, its superiorities will surprise you. And, at the new low price, Kotex is now so inexpensive that it actually costs less than home-made cheese cloth and cotton substitutes.

Buy a box today, at any drug, dry goods or department store ... 45c for 12. It is sold, too, in vending cabinets in rest-rooms by West Disinfecting Co. Kotex Company, 180 North Michigan Ave., Chicago. Kotex Company of Canada, Ltd., 330 Bay St., Toronto, Canada.

KOTEX
The New Sanitary Pad which deodorizes

Kotex advertisement published in *Ladies' Home Journal,* April 1929 (courtesy of the Schlesinger Library, Radcliffe Institute, Harvard University)

Advertisements for menstrual technology were significant in the articulation and shaping of the vision of the modern female body. By the 1920s, manufacturers understood well the importance of advertising in establishing their menstrual products as part of the array of new technologies necessary to produce "modern" womanhood. Manufactured disposable sanitary napkins were first introduced onto the American market in the 1880s and were sold sporadically through mail-order catalogs, via ads tucked in the back margins of women's magazines, and in drugstores. What little publicity these products had gained by 1900 dropped off in the first decades of the twentieth century, and while they appear to have been sold continually in small quantities in drugstores, the market did not really take off until Kotex began its 1921 campaign of intensive national advertising in major magazines.[6]

Kimberly-Clark executives not only had to decide to invest in those advertisements, they had to convince the magazines to accept the ads. Albert Lasker, the advertising executive who carried out this delicate operation on Kimberly-Clark's behalf, relished telling the story of how he convinced Edward Bok, the conservative editor of the eminently respectable *Ladies' Home Journal*, to accept Kotex ads. According to Lasker, Bok had rejected the proposal and was ready to usher Lasker out of his office, when Lasker made a last-ditch appeal: if Bok's secretary thought the ads were inoffensive, would he accept them? Bok agreed and called in his secretary, who turned out to be a "dignified, white-haired lady, seemingly around sixty." To both of the men's surprise, she declared the ads to be in good taste and of great benefit to women.[7] Once Lasker had won a spot for Kotex ads in the well-respected, standard-setting journal, it was easy to buy advertising space in a whole range of magazines and newspapers, and other manufacturers followed suit.

Advertising aimed at consumers was concentrated in middle-class women's monthly magazines such as the *Woman's Home Companion, Good Housekeeping, Ladies' Home Journal, McCall's,* and *Cosmopolitan,* as well as tabloids such as the monthly magazine *True Story* and the weekly newspaper insert *American Weekly.* Menstrual products manufacturers also advertised in more general-interest magazines, such as *Life* and *Look,* and *Hygeia,* a monthly health advice magazine published by the American Medical Association. Advertising was placed in medical and nursing journals and drug retailing journals, urging medical professionals to recommend the products to their patients, and retailers to stock and display the products.[8]

Manufacturers spent a tremendous amount of money on advertising. Tampax spent $100,000 on advertising in its first nine months of existence, a huge proportion of its startup costs. Reflecting later on their first years in business, most of those involved agreed that the key to successfully introducing tampons onto the market was heavy, consistent advertising. By 1941, despite its small size, Tampax was one of the one hundred largest advertisers in the United States. Its two major competitors, the Personal Products Corporation and International Cellucotton, advertised their tampon brands nearly as heavily, although they maintained a greater focus on their sanitary napkin products. Tampax estimated that a well-read woman would see twelve or so ads for Tampax each month, and she would have seen nearly as many for Fibs and Meds tampons, Kotex, and Modess.[9]

This huge volume of advertising was aimed at a predominantly young, white, middle-class and aspiring middle-class female audience. While most magazines do not seem to have collected detailed demographics about their readers, most of those in which tampons and sanitary napkins were advertised touted their readerships as young, female, and modern-minded, with above-average family incomes.[10] In this period, nearly all advertising was aimed only at the top one-third to two-thirds wealthiest consumers, who were assumed to be native-born whites.[11] Women's monthly magazines would have reached a portion of the wealthiest third of the population, and tabloids would have attracted a significant number from the middle third. While this was not as broad a market as menstrual product advertising would target by midcentury, it indicated that menstrual product manufacturers believed that a significant majority of women would have access to disposable menstrual technology, if they chose to stretch their budgets in that direction.

A 1939 Gallup poll indicated that more than two-thirds of Americans may have considered themselves to be in the target audience for a product supporting modern, middle-class self-presentation. Gallup asked Americans, "To what social class in this country do you feel you belong—lower, middle, or upper?" Given these choices, 75 percent considered themselves part of the middle or upper-middle class and 11 percent part of the lower-middle class. Gallup then asked, "To what income class in this country do you feel you belong?" This time, only 41 percent of respondents put themselves squarely in the middle class, 21 percent said they belonged to the lower middle class, and 31 percent said they belonged to the lower class. Coming out of the Depression, the vast majority of Americans already considered themselves to be, in some sense, middle class, but many did not have the income to easily sustain a middle-class lifestyle. Advertisements for menstrual products helped to define and reinforce the standards of bodily practice that would allow someone to define herself as middle class and to communicate this class standing to those with whom she came into contact. They also offered a relatively affordable product to facilitate this self-presentation, which must have been important to many people who perceived themselves socially as middle class but who could not afford to participate in the culture of consumption that was coming to define, in many ways, American social life.[12]

In its first years, Kotex advertising truly appealed to the upper and upper-middle classes with its emphasis on the health and hygienic advantages of Kotex, accompanied by pictures of glamorous women in fur coats, traveling in cars, on trains, and on cruise ships. By the end of the 1920s, Modess competed with pictures of young flappers teaching their mothers how to do the Charleston, ride in motorized boats, and other youthful, less obviously upper-class-bound adventurous activities, in a series of ads titled "Modernizing Mother." Kotex followed Modess's lead, though it continued to target somewhat older women. In the 1930s, sanitary napkin and then tampon advertising converged on a clear vision of "modern" womanhood and the role of menstrual products in realizing it.[13]

The modern, middle-class female body promoted in menstrual product advertising was a well-managed body, which at the same time did not display evidence of its modes of management. A primary requirement of menstrual technology, therefore, was that it not be visible while in use. Early models of Kotex had been undeniably bulky, rectangular blocks of rather stiff Cellucotton wrapped in gauze. Kimberly-Clark quickly changed the design as it received feedback from its customers, rounding and pressing the ends to make them less visible. In the months before Tampax began national advertising, the International Cellucotton Company was bragging that Kotex provided "absolute invisibility—no tiny wrinkles whatsoever. Even the sheerest dress, the closest-fitting gown, reveals no telltale lines." Tampon manufacturers, however, could make these claims somewhat more persuasively, because as they noted in every ad, tampons were "worn internally." A tampon not only hid menstrual blood, it was hidden itself. Wix advertised that its tampons were "utterly invisible," and provided protection "without the slightest tell-tale evidence." Menstrual pad technology, which included a belt and pins to hold the pad in place, could only maintain its invisibility given the assumption that a certain amount of clothing would be worn, for example a slip under a sheer or tight dress. Tampons required much less, and could therefore be worn with fewer and lighter clothes, such as bathing suits. Tampax noted, "There is nothing that can possibly 'show through' a snug swimsuit, whether wet or dry."[14]

Advertisers could not be satisfied, however, in simply demonstrating the discreetness of Kotex or the invisibility of tampons. Insisting on the necessity of invisibility, advertisements promoted women's perception that they were con-

"NEVER MIND, MOTHER - YOU'LL LEARN"

WHAT A SPLENDID game it is, these joyous, fearless, modern girls are teaching mothers— the game of escaping the bondage of old-fashioned ideas and being happily young again.

Middle age is too often resigned to things as they were; youth is resigned to nothing but the best.

Modess has won the universal acceptance of young women simply because it is almost unbelievably better — because it releases them from the unpleasant drudgery of the old way —because it is truly comfortable.

The convincing superiority of Modess is due to a remarkable new substance used for the filler. It is a fluffy gentle mass like cotton, so yielding that irritation is impossible. This filler is amazingly absorbent and is, of course, disposable. Because this filler is so soft, pliant and conforming, the sides of Modess are smoothly rounded and shaped. For added comfort, the gauze is cushioned with a film of cotton.

Modess is deodorizing. Laboratory tests prove it to be more efficient in this respect.

Modess is made in one size only, because its greater efficiency meets all normal requirements without readjusting size of pad. A box lasts longer.

We are sure you will agree with nearly two million women that Modess is finer than anything else you have used. It costs no more than you usually pay. Why not try it?

Johnson & Johnson
NEW BRUNSWICK, N. J., U. S. A.
World's largest makers of surgical dressings, bandages, Red Cross absorbent cotton, etc.

MODERNIZING MOTHER... *Episode Number Five*

Modess
(Pronounced Mō-dess)

SO INFINITELY FINER

An example of Modess's "Modernizing Mother" advertising campaign, from *Ladies' Home Journal,* July 1929 (courtesy of the Schlesinger Library, Radcliffe Institute, Harvard University) stantly being scrutinized for evidence of their menstrual status. One ad headline told women, "Sharp eyes cannot tell with Tampax." It went on, "You need never fear that *anyone* can detect anything if you wear Tampax." Years later, Tampax continued the theme, saying, "In bathing suit wet or dry, you are safe from the most watchful eyes." The scrutinizers were unnamed in tampon ads, but in contemporaneous ads for deodorants and douches, the judgmental observers were boyfriends, husbands, and female peers. For example, a 1936 Mum deodorant ad displayed the headline, "Suppose we listen to the MEN awhile." Readers witnessed men talking about the "unpleasant body odor" of "underarm perspiration." The ad claimed that the quality men look for in women is "daintiness of person," and then busts of men were shown with quotes underneath, such as "Perspiration odor contradicts everything I want to think about a woman" and "I can't forget it—or forgive it—when a woman offends with perspiration odor." Women were threatened with the idea that others might be talking about bad smells emanating from their bodies; however, in the ads for this entire cluster of products, it is clear that this threat was a strategy for convincing women to prevent the possibility of this kind of talk. Women were being encouraged to manage themselves more thoroughly, in ways made newly possible by the advertised technology.[15]

A second requirement of modern menstrual technology was that it be easily hidden before and after its use. One advertisement claimed that Modess, unlike Kotex, worked so well, there was no need to carry a spare. In the ad, Jean sees her friend Sally getting ready for a fancy party. "'Sally!'—she gasped—'You can't go out like that! That valise looks dreadful with your new silver dress! Where's your brocade evening bag?'" Sally eventually goes to the party with the appropriate bag because she is wearing a Modess pad, so she does not need to take an extra one along for "fear of embarrassing accidents." Tampons again had the advantage over sanitary napkins. Tampons were not only invisible in use; they were promoted for their invisibility before and after they were used, as well. Because they took up less space than pads, they could be stored more surreptitiously. Tampax and others advertised that "a month's supply will go into an ordinary purse." Tampax also reminded women that "you can always keep an 'advance supply' in your purse or desk or locker—so inconspicuously!"[16] After they were used, tampons could be disposed of easily, because they could be flushed down the toilet without fear of clogging the pipes. Kotex notoriously made the same claim, despite the fact that its product regularly clogged toilets and caused major plumbing problems.[17] Tampax declared much more convincingly and with a covert reference to the problems created by Kotex, "The whole thing is so compact there is no disposal problem."[18]

Not only blood and menstrual products needed to be hidden; menstrual odor was supposed to be undetectable as well. In 1929, Kotex began including deodorant in its pads and warned in an ad that "whenever women meet the world, they are in danger of offending others at certain times," but if they used Kotex's pads with deodorant, "the last problem in connection with sanitary pads is solved." Tampax touted its product with the claim that "being *worn internally*, it can cause no odor, which eliminates another worry." Meds advertised that there was "no chance of odor (odor forms only on contact with air)." In addition, one of the major selling points of tampons was that a woman wearing a tampon could take a bath or a shower without taking it out. Bathing had long been a middle-class remedy for undesirable body odors, and tampons allowed women to bathe daily during menstruation as during other times of the month, if they were willing to abandon traditional concerns about contact with water during menstruation.[19]

Odor was one of the possible social "offenses" most emphasized in advertising in the 1930s and 1940s.[20] It was the problem that mouthwashes, deodorants, scented soaps, shampoos, douches, suppositories, and underarm dress inserts were designed to control.[21] Concern about odor was already evident at the beginning of the century, but at that point, the focus was on individual comfort at least as much as on others' perceptions. A 1901 ad told its readers than when Rubifoam mouthwash was used, "the results in mouth appearance are enjoyed not by the user alone, but by society as well." By the 1930s, however, the focus was entirely on others' perceptions, and the goal was not simply providing enjoyment but preventing ostracism. A 1938 Mum deodorant ad reminded, "Girls who win men's love keep charming, keep attractive—with Mum . . . Remember—no girl is attractive who isn't dainty."[22]

Advertising was particularly necessary to convince people to manage their bodies with these technologies, because as many ads pointed out, the offender frequently could not detect her own odor. Worse, people as close to her as her own husband would be too embarrassed to tell her that she smelled bad and would simply avoid her. One Zonite douche ad claimed that vaginal odor is *"the most serious deodorization problem any woman has . . . one which you may not suspect."* Another Zonite ad asked, "How can he explain this to his sensitive young wife?" She looks distressed as she watches him head out the door, and he is thinking, "There are some things a husband just can't mention to his wife!" Listerine told stories of women who had lost their lovers because of "halitosis" and sadly watched them marry other people. Advertisers urged women to adopt a host of odor-preventing technologies as a precaution against offenses that they would never know they had committed.[23]

As with menstrual technologies, these technologies aimed at odor prevention were ideally undetectable. Quest deodorant, made by Kotex for use on sanitary napkins, was "unscented, which means it can't 'give itself away,' can't interfere with the fragrance of lovely perfume." Tampons had a special advantage in this area because they were supposed to prevent odor from developing in the first place. "Odor is eliminated, because Tampax prevents its formation." Perhaps this was not enough reassurance for everyone, however; one small brand of tampons, Moderne Women, advertised that it "brings you an important *additional* advantage—*it contains an effective deodorant!* You are *doubly sure* of not offending!"[24]

Menstrual pads and tampons were supposed to be undetectable not only by sharp eyes and noses but also by the wearer herself. Kotex and Modess consistently emphasized how soft and comfortable their products were. A major selling point of tampons was their comfort, described as lacking the sensations normally caused by pads. One of Tampax's first slogans, copied by other companies, was "no pins, no pads, no belts!" Meds elaborated it, saying, "Good-bye hot binding belts, sticky, chafing pads, finger-pricking pins." Tampax claimed dramatically that "chafing, bulkiness, binding become merely a memory of what soon seems a dark era." In fact, the tampon was so well hidden that according to Tampax, "the wearer is completely unconscious of its presence."[25]

The end result of all of this bodily management—which was, itself, undetectable except through absences of blood, odor, or sensation—was the ability to be dainty, poised, at ease, and "modern." These traits were a part of one's "personality," a concept that came into being in the early twentieth century. Personality, a modern replacement for the nineteenth-century concept of "character," was understood as a mixture of mental, moral, and physical characteristics that could be improved in order to get a job, make friends, or win a husband.[26] How a woman looked and smelled was important to her personality, in themselves and in the mental and moral qualities that they helped her project. A Zonite douche ad makes clear how mixed physical and moral traits were in the concept of personality. It began, "If only every young woman could realize from the beginning of her marriage how important vaginal douching often is to intimate feminine cleanliness, health, charm and happiness."[27]

Ads suggested that menstruation could easily undermine important personality traits. A Modess ad illustrates the problematic quality of being ill at ease. It shows a woman who has pulled a friend into a corner at a party and is whispering to her with a frown on her face. The caption says that with Modess, "You needn't be looking in mirrors or asking people 'Am I all right?'" Even if a woman managed to remain "dainty," she was still in danger of revealing that she was menstruating through her manner. Kotex tried to reassure women, asking, "Why be self-conscious! With Kotex your secret is safe!" However, Tampax seemed to suggest that Kotex would not work well enough for this purpose; in one Tampax ad, one woman, talking to another about a mutual friend whom they had persuaded to try Tampax, said, "Well, I hope it'll make her less self-conscious on such days. She always wore such a tell-tale expression." The use of specific menstrual technologies was linked through physical management to mental and moral personality traits such as poise and ease, which were coming to be highly valued by middle-class women and their employers, friends, and husbands.[28]

In the ads, these personal qualities were prerequisites for satisfactory performance of the upper-middle-class activities depicted, such as playing golf and tennis, horseback riding, traveling, vacationing, hiking, skiing, and attending formal evening parties, as well as a few activities that were embraced by college flappers and working girls alike, such as swimming and dancing.[29] These activities were associated with particular social identities. A 1939 Kotex ad featured a "private secretary," the young, employed woman's epitome of the "modern," middle-class identity, saying that she used Kotex because "Looks Count Plenty . . . in this job of mine. A girl must look poised and efficient and that means I must *feel* my best—can't afford to be uncomfortable no matter what!"[30] "College girls" were mentioned in nearly every Tampax ad. These were women who were likely to have already achieved the young, white, and upper-middle class image the ads promoted, and in addition, were those to whom many Americans of all ages had looked since the 1920s as the setters of modern trends in fashion and self-presentation.[31] Consumer research eventually revealed that they actually were some of the most eager adopters of tampon use, and they were set forth as a model of modern thinking for other women. "Young housewives" and nurses were the other groups most frequently mentioned, young housewives for similar reasons as college girls and nurses because of their authority as medical workers. Clearly upper-class identities such as "socialite" and "clubwoman" were mentioned frequently in ads for pads and tampons. Sometimes identity was even vaguer than this, while still maintaining a clear social class. One ad proclaimed, "SHE wears Tampax . . . that socially alert woman whose poise you have admired and perhaps envied."[32]

Some groups of working women besides nurses were also mentioned, including "business women," "office girls," "sales girls," models, and teachers. These young, urban, working women probably did not participate in many of the upper-middle class activities mentioned in the ads, such as golf and tennis. They were more likely to be part of the aspiring middle class and were already blurring class boundaries by working in white-collar jobs. Advertisers wanted to include them as potential customers and worked to sell them an entire image of middle-class womanhood, which included the use of modern menstrual technologies. Tampon manufacturers particularly found themselves having to reach out to the aspiring middle class, to persuade them to use tampons instead of sanitary napkins. One set of ads promoting Meds's low price, equivalent to that of Kotex, showed one woman saying to another, "Once it was a luxury, Dora—But now anybody can afford the modern, *inside* way." The two women are chatting while finishing a game of tennis.[33]

Despite ads' insistence that tampons were for use by "all normal women," not all working women were included in the tampon ads, nor in sanitary napkin ads either.[34] Blue-collar jobs such as working on factory production lines and farming received no mention until the United States entered World War II. At that point, the ads referred to factory work as "war work"; the assumption was that the women targeted by the ads would not have worked in factories before and were doing so only as a patriotic duty. Ads mentioning factory work lasted for less than two years, from June 1942 to April 1944.[35] At that point, blue-collar work became unmentionable once more, and discussion returned primarily to the leisure activities of college girls and young housewives.[36] While menstrual product manufacturers and advertisers were eager to include as many women as possible in the market of the established and aspiring middle class to whom they promoted their products, they defined middle-class identity against blue-collar work. This definition contributed to blurring the line between the established middle class and those who took clerical and sales jobs in the hopes of joining the middle class, but it sent a message to blue-collar workers that no alteration in bodily self-presentation could make them middle class, at least while they were at work. The message to blue-collar workers was instead, perhaps, that adopting a "modern" self-presentation was necessary to attaining the white-collar jobs that would facilitate their entrance into the middle class, or that they would have to become modern through their leisure and consumption patterns rather than at work.

During the 1920s and 1930s, while ads often mentioned "young housewives" in passing, their images focused on the activities of "new women" outside of their housekeeping and family duties. While new women of the 1920s aspired to companionate marriages complete with children after their flapper days were over, their youthful activities seemed to better represent the innovative spirit and new freedom in the public sphere that advertisers wanted readers to associate with their products. By the mid-1940s, however, images of family life at least occasionally appeared in menstrual product ads. Women fed and groomed children, pushed them in strollers, and played with them at the park.³⁷ The emphasis was on activities that mothers did with their children in public, but there was also the suggestion that a mother could use the advertised product to stay cheerful and efficient with her children at home. The woman who used the advertised products to flirt and dance with confidence all month could later use them to enhance her family life with the husband she snared in her youthful socializing. Still, the mother figure so central in advertisements for products from Hostess Cakes to Clipso soap and Esmond blankets was a limited presence in menstrual product advertising.³⁸ Advertisers were careful to balance images of mothering and housekeeping with representations of carefree young women frolicking on the beach or ballroom dancing, generally within the same advertisement.

Menstrual products advertisers in the 1920s through the 1940s presented a vision of modern menstrual management compatible with contemporary trends in sex education, physical education, and industrial health. They depicted active, athletic women who worked and played with aplomb while they had their periods, and who expected scientific and technological innovations to support their efforts. Advertisers drew upon the emerging logic of modern menstrual management, making their appeals on grounds women had already begun to work out in collaboration with educators and health workers. At the same time, ads exaggerated, almost caricatured, the image of the modern woman, and gave it a strong commercial slant. They made much more explicit the relationship between becoming "modern" and maintaining or gaining middle-class status. Ads for menstrual products did not introduce a new idea about modern bodily management, but they did crystallize it, link it strongly to technology purchased in a national mass market, and make its class implications more apparent.³⁹

―⠀⠀―

Beginning in the 1920s, women eagerly adopted disposable sanitary napkins and gradually developed a new set of practices around menstruation not only to manage menstrual blood itself in a new way but also to manage the new technology. Disposable napkins offered a solution to many of women's dilemmas concerning menstrual management, but they also created dilemmas about how to obtain, store, and dispose of the new technology discreetly and without undue anxiety. In keeping with Progressive aspirations, women adopted technology that reduced the manual labor involved in dealing with menstruation but in the process took on a range of new managerial challenges.

Mary Hanson was sure that she began using Kotex shortly after it was introduced in 1921, when she was a teenager. Kotex would have been quite expensive for a young woman working as a telephone operator, at a box of 12 for 60 cents, the equivalent of $5.77 at the end of the century,[40] but Kotex quickly faced cheaper competition. It is likely that she actually bought a knockoff, which could be obtained for less than half the price. "I was tickled to death! [Laughter.] We thought it was wonderful. You would, too, if you had ever worn those other things. It was like putting a harness on you. [Laughter.]"

To Mary, it seemed obvious (as it most likely does to twenty-first century readers) that commercial, disposable pads were far preferable to homemade cloth napkins. This perspective, however, only became taken for granted once "modern" expectations about work and play during menstruation, and modern ideas about disposability, had been established. Disposable products began to be available in drugstores and mail-order catalogs by the late 1880s, but their availability in catalogs dropped off around 1900, and they did not really take off as a product for the masses until the 1920s.[41] Aside from a few discreet descriptions in catalogs and occasional small ads in the backs of women's magazines, there was little advertising, and so it is understandable that they did not develop a large market quickly. But even if disposable pads seemed the obvious choice to women who did notice them, why were their sales not robust enough to at least continue selling them by mail order? Most likely, not enough women thought they were appealing enough to pay for them. Americans had not yet adopted a culture of disposability; reuse was standard practice in all areas of life.[42] Throwing away menstrual pads would have seemed extravagant to anyone but particularly to poorer women, who were most likely to be women who worked throughout the month at physically demanding jobs and therefore would have found disposable pads clearly more practical and comfortable. Women who could afford to buy disposable pads were unlikely to be working women, and for health reasons they were likely to rest at home for at least part of their menstrual periods, at least if they had not yet been swept up in the beginnings of "modern" ideas about menstruation and health. For disposable pads to become as obviously appealing as they were to Mary, women had to become "modern" in their willingness to throw things away, and in their expectations about being able to play and work as usual all month, while meeting their own and their teachers' and employers' standards of efficiency and self-presentation.

Women adopted disposable pads because they seemed clearly more practical, convenient, and comfortable, given modern expectations about activities during menstruation. Some interviewees also indicated awareness that choosing commercial pads was a social and political choice as well as a practical one. Like Mary, Ida Smithson eagerly switched to disposable pads, despite the cost, once she learned about them from other girls at school. "I was raised without a father. My brother used to give us so much every week, so instead of buying something, I bought me a box of Kotex, which I needed. And then had to keep it hid." She never told her mother, a tobacco farmer, that she had decided to use her money to start "being modern, using Kotex, all this kind of stuff." The way women managed menstruation was a clear marker of class status in the rural African American community in which Ida lived, and it seemed important to Ida to adopt "modern" ways in her quest to improve her standing, using Kotex as well as studying hard and graduating from high school at the top of her class. In different circumstances, this trajectory would have led next to college and immediate entry into the middle class, but in the midst of the Depression and without money for tuition, Ida had to resign herself to work as a beautician and housecleaner and help her husband run a small business. She did, however, have the satisfaction of sending her five daughters to college and graduate school.[43]

As Joan Brumberg demonstrates, some of those who rejected the cloth pads their mothers used in favor of disposable pads were the daughters of immigrants. Their mothers did not want to pay for disposable pads, and some mothers who immigrated from Italy also thought that cloth pads were healthier, the gathering of large amounts of blood in a single pad drawing out the rest of the blood that should be discarded.[44] The daughters rejected their mothers' ideas about health and appropriate expenses, insisting that disposable pads be treated as an absolute necessity in the household budget. Brumberg describes this as an important aspect of "Americanization" for the daughters. "Americanization" was certainly an important part of the story, but as interviews conducted for this book demonstrate, it is part of a larger story about an aspiring underclass in the early twentieth-century United States, which included many African Americans and some native-born poor and working-class whites, as well as immigrants. Many young women from these groups aspired to middle-class status, in keeping with Progressive promises of assimilation and advancement, and realized that they would never be able to join the American middle class without adopting the bodily practices and displaying the bodily marks that had come to signify middle-class status.

By the 1940s and 1950s, the use of cloth became a marker of real poverty, and disposable pads came to be regarded as a necessity even for many of those who did not consider themselves to have made it into the middle class. This accompanied similar trends in other areas of material life. In her study of American patterns of household technology and practice, historian Ruth Schwartz Cowan explains that at the beginning of the twentieth century, the differences in material life between those who were comfortable and those who were struggling was stark, and considerably more than half the population did not live comfortably. By 1950, a standard of living that would have seemed fairly luxurious at the beginning of the century—including a dwelling with a complete bathroom, a complete kitchen, and central heat, and a yearly set of new clothes—was attained by almost three-quarters of the population.[45] This shift was reflected in the lives of those interviewed for this book. Rachel Cohen knew that her mother, born in 1904, had used rags when she was younger, but by the time Rachel came of age in the 1940s, "even though . . . we didn't have discretionary money in my family, we bought sanitary napkins."

Even those living in more marginal economic circumstances were likely to do what they could to buy disposable napkins. Mara Ozols, thrilled to discover disposable pads when she arrived in the United States in 1950, stretched her budget to include them. "I was buying napkins here . . . I say, 'Oh, look at that! Don't have to make, and don't have to wash! . . . That's enough [of rags]. That's enough.' And still, you know, was no money too for napkins, but you wear 'til the last drop! [Laughter.] But I buy napkins." She ran the family farm and worked night shifts at a factory to support her four children and disabled husband but did not consider saving money by using cloth pads.

Cloth pads were still imaginable in some communities, but having to use them was the worst-case scenario. Roberta Cummings Brown, almost twenty years younger than Ida Smithson, recalled learning from her classmates, "worse comes to worst, because there were some people who were poor, and they said, if you don't have anything, then you just take some rags and fold those up. And then they also showed how you take a lot of toilet tissue, a lot of napkins and fold those up and make one." Liza O'Malley, the daughter of blue-collar workers, explained, "not in my crowd, but some in my generation that I went to school with that were poorer would wear cloth. And they used to call it "the rag," and that always made me very uncomfortable . . . They talked about them because they couldn't afford them, so they'd have to go home and bleach them." This lower-class way of talking about and managing menstruation made Liza uncomfortable because it "contributed to making [menstruation] feel dirty" to her. By the 1940s and 1950s, the line was drawn between the truly poor and the ever-broadening middle and aspiring middle class, who shared a specific set of routine intimate practices.

When women switched to disposable pads, most of them bought Kotex. Kimberly-Clark dominated the market with the Kotex brand, maintaining more than a 70 percent market share through 1953, despite competition from cut-rate brands as well as Johnson and Johnson's widely advertised Modess brand.[46] A few interviewees who were born before 1940 could remember other brand names, but all said they used Kotex. Jill Okada Sun mused, "I guess the Kotex was a brand that we used. In fact, that's probably one of the major ones that they sold in those days, in the forties, so I think I just remember this blue box that said Kotex . . . I think by association you just depend on buying that product."

Once the older generations adopted Kotex, their teenage daughters usually simply used the supplies that their mothers kept in the house. Ida Smithson had hidden her Kotex from her own mother, who used rags, but she had specific reasons she wanted her daughters to share her supply. "I would have them; they had to tell me when they needed some. 'Mama, I need some of those,' and so I bought them a box, to make sure that I knew that they were having their period! [Laughter.] They had confidence in me, and so they would tell me when they needed them."

Kotex was an unusual mass-market product in the new culture of consumption, in that while it was advertised aggressively and its use signified class standing, no one wanted to consume it or even retail it conspicuously. As Sharra Vostral has pointed out, Kimberly-Clark tried, in its early advertising, to convince its customers that asking for Kotex would be less embarrassing than asking for sanitary napkins, and that the blue box was discreet enough to satisfy the most nervous customer. However, *Kotex* rapidly became synonymous with *sanitary napkins*, and Kotex's blue box rapidly became recognizable, forcing women and clerks to find other ways to avoid embarrassing exchanges.[47] Ida Smithson recalled how much she disliked going to the store to buy Kotex. "I was buying from the dime stores. And you would shame, when you went to get it. You looked around, to see, is anybody watching you. [Laughter.] . . . They would say, 'Hah, look! Look what she's buying!' . . . Some people stand and watches. It's terrible. [Laughter.]" She explained how she and other women would handle this uncomfortable situation, wondering, now, why they had done it that way.

You had to ask for it. They didn't have it out there. Because somebody would say, they would say, "Where's those things?" [Hearty laughter.] Didn't give the name. They knew what you were talking about when you said "those things," and they would reach behind the counter. Kept them behind the counter! But you know, that's strange. Why would they hide it?

Once she had been handed the box, "I would just take a look around, and see if anybody sees you ask for it, and you kind of slip it in the bag, something like that."

Feminist scholars have pointed out, rightly, that menstrual products manufacturers have contributed greatly, through their advertising and other practices, to the sense in American culture that menstruation is embarrassing and must be kept hidden.[48] It is true that manufacturers aimed most of their product development at creating products that hid menstruation as thoroughly as possible, and then, in touting those benefits of the products, promoted the idea that menstruation should be kept secret. However, manufacturers were not necessarily pleased at the direction their industry took. It was difficult to promote and sell products that people were embarrassed to be seen buying and that retailers were uncomfortable promoting locally. In addition, it was difficult to move the product out of the "commodity" category and turn it into a product with a larger profit margin, if its consumption could never be conspicuous. Manufacturers often only reluctantly gave in to demands of secrecy, or at least modesty, from women, retailers, and advertising venues.

Besides encouraging women, in its 1920s advertising, to ask for Kotex by name, Kimberly-Clark begged and prodded druggists to actually display Kotex boxes and promotional materials. Representatives visited stores, awarding cash prizes for the best displays in annual contests. In the trade press, they advertised with headlines such as, "Kotex, Don't Hide It!" All manufacturers competed to get large and prominent retail space for their products, but Kotex, Modess, and other sanitary pad manufacturers particularly faced an uphill battle. By 1924, as an attempt to compromise with retailers' demands for greater discretion, Kotex promoted a ready-wrapped package to retailers, so that individual druggists did not have to wrap the Kotex boxes in brown paper themselves. These prewrapped packages were supposed to be placed in a display with a single unwrapped package on top, rather than in the back room or behind a counter, so that customers could pick up the boxes themselves. They advertised to the trade that this innovation "makes Kotex as easy to buy as soap. It removes the only obstacle to drug store sales—that of asking a man clerk for it, especially within hearing of other customers." Apparently, the brown paper did not hide enough; in 1928, the Kotex package was further modified to tone down the lettering so that it would not show through the brown paper wrapping.[49]

Despite all of this care taken to help retailers wrap packages, Kimberly-Clark still hoped that retailers could be convinced to create eye-catching Kotex displays, if only the packaging were nice enough. In 1934, they changed the package design, to two-tone blue for regular size, green for junior size, and brown for super size. "The combination of colors on display was very attractive but it was a very loud package, designed for maximum display value."[50] Given interviewees' reports, though, it is unlikely that the bright new packages ever saw the light of day. Rachel Cohen recalled the process of buying Kotex in 1930s Denver, describing it in the typical setting of a corner drugstore.

> Something else about that whole era is, to buy sanitary napkins, you went into a drugstore. Now, those days they didn't have big drugstores like Longs and Payless and all those. They just had little family drugstores. And the drugstore was literally that. There was a pharmacist, who was usually a member of the family. I had a girlfriend whose family had a drugstore. The father was at the counter, and the mother was the pharmacist, actually. In those days a lot of medicines were made up with a mortar and pestle. And they sold sanitary napkins and they also had a little candy counter, where they sold candy. And Kotex was always prewrapped, when they got a shipment of Kotex. They used to prepackage them in brown paper and stack them in the back, you know, in the store but in the back, and a woman would come in and she wouldn't say anything [like], "So I want a box of Kotex." She'd go over and get the package and just lay it on the counter and they'd say, you know, seventy-five cents or whatever it was. But nobody used the words. So that's the way you got Kotex.

Frustrated with the conventions surrounding Kotex purchasing, Kimberly-Clark continued to make sporadic attempts to get rid of the brown paper wrapping, which cost extra money to provide and hampered product promotion. Still, they seemed willing to promote the wrapping if it was in the service of a worthy cause, such as convincing teenage girls that it was OK to ask a male clerk for Kotex. In their 1940 pamphlet for girls, *As One Girl to Another*, Kimberly-Clark made a claim that must have raised a few teenage eyebrows:

> you need never feel the least embarrassed to ask for a box of Kotex in a store . . . even if it's a tall, young red-haired lad on the other side of the counter! He'll give you a

box already wrapped without batting an eye. He is so used to selling Kotex it's just all in the day's work to him. Like selling tooth paste or talcum powder. Honestly!⁵¹

Prime competitor Modess, focused more than Kotex on young customers, made its paper wrapping the centerpiece of its advertising to both women and retailers when it moved into supermarkets in the late 1940s, and Kimberly-Clark reluctantly followed suit. Modess told women that with a new box shape and an overwrap, "we'll help you keep your secret." When Kotex did not prewrap its boxes in 1951, distributors opened the cases and wrapped the boxes before selling them to retailers.⁵²

Kimberly-Clark kept trying, making a concerted effort again in 1954, changing its package from blue to "soft silver grey," after conducting consumer research that "showed that this package was less conspicuous and preferred by consumers over wrapped packages by Modess . . . more feminine and more attractive . . . designed to help us promote the fact that Kotex should be merchandised and marketed like any other commodity item." After creating the new packages, Kimberly-Clark advertised aggressively to the trade, to convince retailers that these new packages deserved to be displayed prominently. Engaging the same old debate about the propriety of displaying sanitary napkins, they argued that first, Kotex was a commodity just like soap, deodorant, and toilet tissue; second, "Buying Kotex is a symbol of cleanliness;" third, "Kotex is advertised just like any other product;" and fourth, women wanted to know they are buying Kotex, and not a substitute. They concluded that retailers should position boxes to make them easy to see and easy to reach.⁵³

Apparently, all of these efforts were unsuccessful. The practice of wrapping Kotex in brown paper and hiding it behind a counter continued through the early 1960s, prompting Margaret Olsen, who came of age in the late 1950s, to think at the time, " 'Wow, what is that all about?' Like it was a disgrace, or something."

In December 1958, Kotex executives gathered at a brainstorming meeting to discuss the future of the brand's packaging and spent a lot of energy debating what to do about the continuing wrapping problem. The most recent in a series of consumer surveys they had commissioned showed that 14 to 17 percent of women still felt embarrassed buying sanitary napkins. The all-male group spent a lot of time debating about why these women were embarrassed. One suggested that "women are embarrassed because they think men are embarrassed. Between women there is no embarrassment." Another disagreed, arguing that the 14 to 17 percent of women who were still embarrassed "seem to feel that the other 80 percent are embarrassed so they remain embarrassed." Others argued that it was a result of something more deeply psychological, "because it has to do with the sexual parts of the body" or, more vaguely, because of the long history of menstrual taboos.

Several of the executives gathered realized, along the lines of Margaret Olsen's observation, that Kimberly-Clark's practices might have had something to do with the embarrassment. "To some extent, I think, we lead the women to feeling that packaging is related to this whole subject by deliberately hiding it, making it something that has a social stigma attached to it where it is intimately related to them[,] whereas we should take it out of that intimate relationship." Another noted that this cycle of women feeling embarrassed and Kimberly-Clark trying to assuage embarrassment through brown wrapping had created problems for customers and the company. "I think we're enslaving women with this thing. We've created a monster that we can't get rid of . . . you have to keep changing the package every three months because pretty soon this has a symbolic status. Anything wrapped in a plain bag has to be sanitary napkins. It can't be anything else. You're in the trap of an eternity of package change if you're going to try to hide what's in there."[54]

This executive noted the cycle but perhaps failed to notice that just as much as Kimberly-Clark's practices "enslaved" women, women's discourses and practices surrounding menstruation also constrained Kimberly-Clark. The company did not wrap the boxes simply because its executives believed that buying Kotex was embarrassing but because women had told them for forty years that buying menstrual products was embarrassing and that they were more likely to buy a product that was wrapped in plain paper. Pressure to hide menstrual technology was strong even before manufacturers promised to sell women new, disposable technology that was supposedly easier to hide, at least while it was being worn. The new vision of the well-managed, modern body promoted by manufacturers did not include a mandate to freely discuss menstruation in public or display menstrual products, except insofar as these practices were considered part of a more "scientific," rational, matter-of-fact way of thinking about menstruation. It would take forty years of coaxing and negotiation for women, manufacturers, and retailers to firmly establish an acceptable public space for acknowledging menstruation in support of modern menstrual management.

During the 1958 meeting, after a long discussion of all of the negative attitudes and feelings associated with menstruation, and how this negativity hampered product promotion and sales, the group started trying to come up with ways to give menstruation a positive valence, through connecting it to femininity and pride in womanhood. One participant in the discussion declared, "This is probably a real key to this thing. That really what we want to do is instead of going along with this social attitude toward this thing which has prevailed for so many centuries, let's change the damned attitude."[55]

While this approach to the problem may have seemed like a revelation to this particular executive, Kimberly-Clark had in fact been trying to change attitudes toward menstruation from its first advertising campaign, and invested a great deal in its education department, producing materials designed to influence girls' attitudes toward menstruation from their first periods. Discussing the problem of women still being reluctant to talk about menstruation, the executives debated whether the growing openness in communicating about menstruation they observed was due to Kimberly-Clark's efforts or not. They agreed, in any case, that advertising copy had grown much franker between 1921 and 1958, and magazines and newspapers had allowed them to create and promote a much broader published discourse about menstruation. Finally, by 1961, enough women felt comfortable publicly purchasing menstrual products that Kimberly-Clark and other manufacturers could sell sanitary napkins in boxes that proclaimed brand names, without wrapping them in brown paper before selling them or hiding them behind a counter.

———

This greater acceptance of public display and purchase of menstrual products opened the way for some women to begin asking their husbands for practical assistance with management, just as they began to ask for sympathy and understanding with regard to menstrual discomfort. Interviewees recalled that in the 1960s and 1970s, it was not unheard of for a wife to send her husband to buy menstrual products. Peter Jefferson found it perfectly reasonable that his wife would occasionally send him out for supplies. "A couple of times where we would be somewhere, like out of town or at the beach or something, and it's like, 'oops, I forgot to pack this, and I need it.' And you get this command that, 'listen, my period is starting and you need to go get me something.' And, you know, so you jump in the car and go get it." He did not expect it to be a regular part of his shopping, but it did not surprise him that his wife would ask him to do this.

Born a little earlier, Joseph Wong had done the same but felt less comfortable about it. "That's like carrying a purse down the street. [Laughter.] 'Oh, no! All right, all right.' And even if that's all we needed, I'd get milk, or something else to go around it." Men had not been totally exempt from these duties previously; Mary Hanson recalled a time when she was working at a drugstore in the 1930s when a man came in to "get something for his wife, some Kotex, and I almost dropped dead. Imagine, from a man, too?" She told the story in order to show that "they were getting a little more open about things," but clearly this was unusual given how memorable and worthy of exclamation it still was sixty years after the fact. In contrast, for husbands in the 1960s and 1970s, it could be routine if somewhat embarrassing.

Kotex and other menstrual pad and tampon manufacturers tried to make their products available outside of drugstores and supermarkets as well, through vending machines placed in restaurants, hotels, and eventually schools and workplaces. In certain locations, women could rely upon finding vending machines; for example, in 1926, Metropolitan Life Insurance's headquarters contained fifteen vending machines, from which women employees bought an average of 4,500 sanitary napkins per month.[56] However, vending machines were certainly not ubiquitous, even in locations catering to middle-class women, and poorer women were unlikely to have encountered them earlier in the century. Ida Smithson laughed, when asked if she had a Kotex vending machine in her high school. "Shoot, no! [Laughter.] If we had toilet tissue, we were doing fine!" Since public bathrooms could not be counted upon to contain vending machines, women did not rely on vending machines to buy Kotex when they already had their periods, instead using them only for emergencies. They were a less than perfect solution, since, as Mimi Cummings Wood pointed out, they did not come with a belt, so women would use the pins they came with to pin them to their underwear, which was not very comfortable. Vending machines remained marginal, and women bought their supplies from more conventional retail outlets: primarily drugstores in the beginning, gradually switching to supermarkets starting in the late 1930s.[57]

Even when the pads women bought were not actually Kotex brand, women and drugstore clerks called them Kotex and often understood that to be a generic term for sanitary napkins. Kimberly-Clark fought this trend in the trade and popular presses. Like many corporations in the early part of the century, they were educating consumers and retailers to appreciate brand names and learning how to protect their brand against competitors who would infringe upon it.[58] The company ran trade ads telling retailers that if they substituted cheap alternatives for Kotex in their brown-wrapped packages, customers would be angry on unwrapping them at home. They would be too embarrassed to complain and instead would silently give all their business to a more trustworthy retailer. At the same time, Kimberly-Clark told women not to accept substitutions; if they were too embarrassed to complain to a retailer who sold them a cheap alternative and called it Kotex, they could complain directly to Kimberly-Clark. Ironically, Kimberly-Clark may have brought some of this trouble on itself, since their advertisements explicitly taught women and retailers to substitute the word *Kotex* for *sanitary napkin*. They had advertised that women who were too embarrassed to ask for sanitary napkins could simply ask for Kotex, the clerk would understand this code word, and the transaction between customer and clerk could be completed without embarrassment for anyone. Quickly, the two terms became so identified with each other that *Kotex* was no longer a useful euphemism for drugstore customers and clerks, but it was adopted by women in settings where they were more willing to mention sanitary napkins.

Kotex dominated the disposable pad market partly because Kimberly-Clark advertised and marketed the brand aggressively to consumers and retailers but also partly because competing products could seldom claim advantages over Kotex. Kotex was made from Cellucotton, a patented, highly absorbent material that was more effective than cotton, if somewhat rougher and less comfortable to wear. Competitors, such as Modess, sometimes advertised that their products were softer, but they generally resembled Kotex in terms of shape and bulkiness. It was not so easy to tell the different brands apart, and the slight differences did not seem to add up in favor of any of Kotex's competitors. In a 1926 survey for Johnson and Johnson, makers of Modess sanitary napkins, Lillian Gilbreth found that the college and working women she interviewed generally bought whatever was most convenient to obtain, and Kotex was much more widely distributed than other brands.[59] Women saw no reason to go out of their way and risk extra publicity for their purchases to seek out alternatives that just were not that different.

Kotex in its early versions had other disadvantages besides its bulkiness, roughness, and one-size-fits-all design. As Florence Wu explained, the pads were not lined, so the blood "just went right through it. The gauze kind, and that was it. Nobody ever thought of putting a piece of plastic underneath it. Or putting adhesive on it. It was just *so* uncomfortable. Gee, it was awful! [Laughter.]" Kimberly-Clark was aware of the various problems with their product, since they regularly surveyed customers and asked small groups of women, often factory employees, to test new versions of products and report their experiences in detail. Kimberly-Clark engineers experimented with adding various existing materials, such as cotton, which were commonly used in making menstrual pads, and tried many variations in arranging the materials. They tried to make the pads softer, shaped to conform to the body a bit better, and more evenly absorbent. Overall, the engineers' fiddling slowly improved the product, but occasionally a change caused much bigger problems than it solved. In 1936, they added a wadding side strip to reduce stains, with a layer of cotton on it to make it soft, and the cotton caused unforeseen problems. The innovation "was discontinued . . . because the cotton was too slippery and the wrappers failed to hold together, causing the center of the pad to drop out when worn."[60] It can easily be imagined how upsetting that problem was to Kotex's customers.

Women sometimes did some engineering themselves, adjusting the pads to make them work better or conform better to their particular needs. Gilbreth found that the majority of the college and working women she interviewed modified Kotex before using them. In the 1920s, the pads were simply made of layers of Cellucotton wadding wrapped in gauze, so the gauze could be unwrapped and the wadding trimmed and rearranged to suit a particular woman's body and the extent of her menstrual flow on a given day. At the time, Kotex advertised the fact that it could be easily modified as an advantage of its products over competitors' products. As Susan Strasser has pointed out, this kind of adjustment work would have felt familiar and natural to women in the 1920s, who were only just beginning to use ready-made products, but it would rapidly become an unfamiliar practice as the drive for the convenience of prefabrication came to dominate American consumer culture.[61] Kimberly-Clark's engineers fairly quickly incorporated into the Kotex design some of women's common modifications, including rounding the corners of the pad to make it less obtrusive. More slowly, they worked to design an interior wadding that directed and captured the blood flow by more effective means than simply piling up more wadding near the vaginal opening. These new designs made Kotex into a more effective but less flexible product; instead of adjusting the technology to fit their bodies, women found themselves adjusting their bodily practices to fit the technology.

By the 1950s, interviewees were not re-engineering the Kotex itself but had developed other techniques for ameliorating its continued deficiencies. Roberta Cummings Brown explained, "we made liners, we were pretty innovative . . . As a double precaution, we would take toilet tissue and fold it over, about like the size of a sanitary napkin and then sort of like line your panties, then put this down, and then the sanitary pad." Christina Donvito used multiple Kotex pads, though she was not really satisfied with that solution. "I would just double and triple, and whatever. It wasn't very comfortable, no. I always hated that time. Always uncomfortable."

Along with Kotex, women used belts to hold the pads in place. Using belts with Kotex required some skill, more so than with the self-adhesive pads of the late twentieth century. Interestingly, a number of women noted having to learn how to use belts with Kotex correctly, while those who had used belts to hold cloth pads thought that it seemed obvious how to use them. Younger women were less accustomed to this kind of technological improvisation, and as the technology became less flexible, there was less variation in bodily practices that effectively complemented the technology. Roberta Cummings Brown described the difficulties of learning to use Kotex and a belt without help from her mother and remembered her high school health teacher later providing great assistance.

> In high school in the health class . . . they showed you how to put them on . . . They told you how to pull the wings all the way through, and get that even . . . I had been doing some things wrong, you know, it's like trial-and-error, and sometimes it would be very uncomfortable because I couldn't get the strings—that was during the time when you had the belt . . . It was an elastic belt that, first you stepped in it, and then you pulled through the little wings on the sanitary napkins. Remember, there were little wings, that were longer than they are now, that you pulled through the belt, and it had a pin. And then you pin that to, there was a little piece . . . if you got it on correctly, it stayed, but if you didn't, then you messed up your clothes . . . This was a pack of things that was bought at the drugstore, usually, and it came with the belt and two safety pins. And you pinned that, and if you didn't, and if you moved around and were athletic at all, then it moved, so therefore you messed up your clothes, because it just wasn't stable.

Of course, it was not just a matter of getting the belt and Kotex adjusted properly to catch the menstrual flow; clothing had to be adjusted as well, to hide the belt. Roberta remembered ruefully the consequences of wearing the wrong clothes.

> The belt—it was made out of elastic, and then if you had on anything really thin, you could see through it, you know, you could see this little string go in. And anybody, especially in high school, boys in particular would look for this belt, and then they would say horrible things, like "Oh, she's wearing the rag." [Rueful laughter.]

Hiding Kotex under clothes was easier than hiding cloth, but at the same time as the introduction of Kotex came more fitted clothing styles and less layered outfits, increasing the challenge of keeping menstrual technology hidden. Kimberly-Clark dropped its "Can't Chafe, Can't Fail, Can't Show" slogan from its advertising in 1938 because company leaders feared being targeted under new Federal Trade Commission regulations forbidding false claims in advertising.[62]

All of this technology needed to be stored somewhere, and because Kotex was

Easy Disposal

The illustrations on this and the opposite page show you, at a glance, how to dispose of Kotex.

Open the gauze bandage to permit the inside filler to drop; then flush it away.

Then, holding the bandage by one end, snip off one-third of it with pair of shears.

If these directions are followed, Kotex will not clog the pipes as any other pad will.

How to Fasten Kotex

The gauze used in Kotex is full length. This abundance of material has been provided so as to allow for folding before pinning. If the gauze is pinned by the ends without folding, the tabs will be too long and might pull out.

The proper way to pin Kotex is first to fold

Convenience

Read carefully these instructions for disposal of Kotex. If Kotex offered no other advantages than simply this ease of disposal, women would say that it was important enough.

Now Kotex pads are made odorproof by a new process. Each Kotex pad is permeated with an odorless disinfectant.

Then snip once again—this time half of the remaining length of the bandage. Flush bowl.

Finally, having cut the bandage into 3 pieces, drop the piece you are holding and flush again.

in the two corners of each tab, as shown, then fold each end over two or more times and then pin. If pinned in this manner, Kotex will fit snugly and securely; the bandage will remain flat and comfortable.

The longer tab is intended for the back; the open side, where the gauze folds, is the outside.

8   9

Kotex disposal and fastening instructions from Kimberly-Clark's "Personal Hygiene for Women," 1928 (courtesy of Museum of Menstruation, www.mum.org)

unambiguously menstrual technology, women felt it needed to be kept hidden. For the 90 percent of homes that did not have flush toilets in 1910, and the 24 percent that still did not have them in 1950, bathroom storage of menstrual technology was not an obvious option.[63] While families eagerly installed modern bathrooms as a core component of achieving a middle class standard of living, the "modern" model bathroom designed in the early twentieth century was all about minimalist, clean lines and washable enamel surfaces.[64] Storage cabinets were not a part of this standard design, so there was no discreet place to keep pads within the bathroom itself. Discretion was considered necessary because all members of the family, including men, shared one bathroom. Most women interviewees born before 1940 kept Kotex in their bedrooms. African American women reported keeping their supplies behind or under the bed, one of the few places in their homes they considered to be appropriately private and hidden. This space was not necessarily kept private from all the members of the household, just the men. Presumably, since it would be unusual for men to be doing housecleaning, it was hard to imagine why they would possibly look under or behind the beds. Sisters Roberta Cummings Brown and Mimi Cummings Wood kept a supply in the bedroom they shared, but they knew that if they ran out, they could go to the box behind their mother's bed. Christina Donvito felt that anyplace in her parents' room was unlikely to be sufficiently hidden from her father, so she proposed a different solution.

> We had one little secret spot in our closet. Daddy didn't come in the girls' closet. We always had one little spot . . . we told Mum, "You put yours in with us; this is where they are." We had a plastic container, with a cover, and we put the whole box right in the plastic container . . . It was all hidden, and you know, Mom used to have some different nice little . . . [sachets], so, like a girls' bedroom smell, you know? So it worked ok; it was all very nice.

This kind of hiding was apparently quite effective. Mike Ozols guessed that his mother had used cloth pads instead of Kotex because he never saw her supplies around the bathroom. In fact, Mara Ozols did use Kotex and must have stored it in her bedroom, the only private space she maintained with a large family in a small farmhouse.

In the second half of the century, as the bathrooms in new homes began to be built with storage cabinets under the sink, the standard storage location for sanitary pads and tampons gradually shifted to the bathroom. This meant that supplies were much more accessible for the women using them, but it also meant that visitors, men, and children were more likely to come across menstrual supplies. The architectural and spatial infrastructure to support modern menstrual management kept things "hidden," but hiding places were (necessarily) predictable in order to be truly useful. These hiding places included brown-paper-wrapped packages and bathroom cabinets. Allowing things to be visible as long as they were in sanctioned spaces was an important part of rationalizing care of the body.

For women visitors, the routine practice of keeping menstrual supplies under the bathroom sink could be a means of preserving discretion about their periods. Barbara Ricci was sensitive to this. "I like to have them in all my bathrooms, for company. Just in case. Because I've been in situations where I've needed one, and was in their bathroom, and there was no Kotex in any of the cabinets. And so I've always been aware, to have them there, so no one would ever have to ask me." Marina Cabral, who immigrated to New England from Cape Verde as a child, appreciated this standardization of storage as well.

> It's always been that spot. It's always been there. In the back. I actually go to people's house, and I know that it's there. I remember, I went to my cousin's house, and I saw my period, and I was like, "Oh, my gosh, I'm not home!" And I went right under there, and there it was, and I was like, "Oh, good, thank God, everybody," most people, "has the same."

This standardization of storage in a coed space meant that men were also likely to come across menstrual supplies. Mike Ozols, who never saw his mother's supplies when he was growing up, was aware of his daughter's menarche because "there was more tube boxes, tampons, hanging around instead of one. As far as that goes, she didn't ask me, I didn't talk about it with her." Peter Jefferson made fun of any man who would consider objecting to menstrual product vending machines in coed college bathrooms, comparing it to the situation in his house. "There'll probably be one in a million, some guy that's like, '[silly voice] I'm offended by having to look at Kotex and tampons.' And it's like, 'God, wait 'til you get married.' [Laughter.]"

Children were also much more likely to encounter menstrual supplies under the bathroom sink, often before they had learned about menstruation in an educational context. Adam Chiang, born in the early 1970s, remembered playing with his brother in the bathroom when he was around seven or eight years old. They found a box of Kotex and stuck them to the walls. Another time, they discovered tampons and played with them by "running them under the sink, and doing crazy stuff with them . . . We really didn't know what they were. We didn't have any taboos about it."

Kotex also needed to be discreetly carried along to school and work, another new managerial challenge. Some women could use a pad for an entire day without changing, but many found they needed a spare. As Ida Smithson noted, most girls in her generation did not carry pocketbooks or even book bags to school, so it was difficult to figure out where to carry a spare pad. Rachel Cohen recalled this being an extremely distressing problem, especially since she got her period before most of the girls in her class and suffered nasty teasing.

> It was considered to be very disgusting, and you were ashamed of having all that blood run out of you and having those smelly napkins and to dispose of them. I mean, if you were in school, you obviously had to carry a

spare one with you ... so you had to kind of hide it in your notebook or whatever it was, and carry it around with you.

As they got older, women in this generation began to carry pocketbooks, but the styles did not always accommodate a large bundle of menstrual equipment. Samantha Fried solved the problem for herself by ignoring mainstream fashion.

> I had a handbag from the Village that was like—well, now they're popular—but at the time they were only made in the Village. It looked like a horse's nosebag, those huge kinds of things. So you could put things in a paper bag, or whatever. Plastic was not so readily available in those days. You could put them away in your bag and not have things inside. It was much more a problem to dispose of the used napkins. There just were no provisions in the restrooms!

Girls growing up in the 1950s and 1960s began carrying pocketbooks as young teenagers, which tempered the problem but did not solve it completely. Margaret Olsen vividly recalled the kind of teasing that pocketbooks could generate.

> I can remember in junior high, one day on the playground, one of the boys grabbed one of my friends' pocketbooks. And the boys went running with the bag, and she was hysterical because she had a pad in there. "Oh, my God, they're going to see it, they're going to see it!" We were all hysterical. I remember thinking, I don't think I verbalized it, "Is that why he took her bag? Does he want to see what's in there? How does he know?" I don't remember what the outcome was; I just remember that moment. That being *such* a big thing, especially at that horrible age, at junior high.

And pocketbook carrying was not always practical. "I remember around proms, you'd always be worried, what are you going to do, you just had that little pocketbook, that little beaded satin bag, [and you'd think,] 'What am I going to do if I had my period,' and just hoping that it was going to not hit on that time."

If obtaining and carrying supplies could be tricky, disposal often created bigger difficulties. Kotex and other brands advertised at least through the 1920s that it was possible to take apart a used pad and flush it down the toilet, but in practice, it was difficult and messy to take it apart. Rachel Cohen described the process with distaste.

> You had to peel off the cheesecloth to flush them ... and sometimes it was stuck, you know from the dried blood or something. But you had to take this whole thing apart to flush one part and then another, and then before the toilet got flushed, you walked out of there and pretended you didn't know anything happened. I mean, it was a real ugly mess.

Quickly, as Mimi Cummings Wood recalled, signs were put up telling women not to put pads in the toilet. She learned that pads could not be disposed of in the toilet, particularly in rural plumbing systems that did not include a public water system and sewer, by reading the signs forbidding it at school and church. Lillian Gilbreth, in her report to Johnson & Johnson, had taken manufacturers to task for falsely advertising the flushability of their products, but it took a number of years for companies to stop making these claims and probably longer for women to stop trying to flush them, given the long-term lack of special disposal containers in women's public bathrooms.[65] At home, of course, it was a different story; as Christina Donvito noted, it would be clear who, and what, had clogged the toilet once the man of the house had successfully unclogged it, and that situation would be too humiliating to risk.

If Kotex could not be flushed down the toilet, and it could not be washed like cloth pads, women had to find other ways to dispose of it. Women born before 1940 developed elaborate ways of wrapping used Kotex so that it would not be seen or smelled, setting the trend for the following generation as well. They carried newspaper, paper bags, plastic bags, or some combination into the bathroom with them, carefully wrapped up the used Kotex, and either left the little bundle in the bathroom trashcan or, sometimes in rural areas, brought it immediately to an outside drum to be burned. Women learned how to dispose of Kotex in this way either by direct instruction from mothers, sisters, or friends, or from observing, as Roberta Cummings Brown put it, "You know, the little ball in the trash can. The ball is very obvious, because we all knew how to make it the same way. [Laughter.]" Peter Jefferson confirmed that for all of women's efforts to keep evidence of menstruation hidden from male family members, the balls in the trash can gave it away. "My mother is old school, and she's a Kotex lady. She always had, it was so cute, because she had these neat little packages that she would make. [Laughter.] And like, 'Oh, OK! I know what's in there!' [Laughter.]" Standardized technology and shared ideas about the need for discretion promoted the development of standard practices, which subverted any real secrecy. Properly discarded sanitary napkins, like products being sold in stores or hidden in bathroom cabinets, became an acceptable public sign of menstruation, so long as they were kept within strictly defined bounds of propriety and good management.

Mimi Cummings Wood pointed out that although it was clear that used Kotex needed to be wrapped up, there were better and worse ways to do it. She was pleased to learn a particularly tidy method when she was boarding with a family in a nearby city during her first teaching job. "There was a young lady maybe a couple years older than I was, that lived in that household. And so she just showed me the neatest way to fold up the sanitary pad and wrap it up, so that there was no exposure. And then put it in a bag and discard it like that." While the goal of disposing of pads without revealing any blood or odor was clear to everyone, the practical aspect of how to achieve that goal effectively and reliably was not always obvious, and explicit coaching from someone more sophisticated or experienced could be reassuring.

By the 1950s and 1960s, schools and other public buildings began to make special provisions for menstrual pad disposal. Laurie Wilson remembered that her public high school provided vending machines and little bags in which used pads could be wrapped and thrown in the regular trash can. Even better, women thought, were special receptacles placed in the stalls so that pads did not need to be surreptitiously carried out of the stall into the main area of the bathroom. Bags or special receptacles in bathroom stalls were by no means ubiquitous by the end of the century, though middle-aged and young women had learned to expect them and expressed annoyance when they were absent. In some situations, their absence could create real difficulties. Mimi Cummings Wood taught first grade and used the same bathroom that her students used, which was in the classroom. Dealing with used pads was "a challenge! Tell you what, because you couldn't put it in a trashcan. So what I would do, I would put it in my purse and bring it home." That was definitely a less than ideal solution.

Another aspect of the modern management of menstruation is the practices designed to prevent the odor of menstrual blood from becoming apparent, especially to others. Preventing odor became particularly urgent in the twentieth century, as bodily signifiers of middle-class status became more desirable and necessary for social and economic advancement, and as products for preventing or ameliorating body odors proliferated. Odor was a particularly tricky problem, because washing to prevent it conflicted with long-standing beliefs in most of the cultures from which Americans emigrated that contact with water, especially cold water, during menstruation could be harmful to health. This popular belief was affirmed by doctors' recommendations. Through the mid-twentieth century, many physicians continued to advise women not to swim or take cold baths or showers, and to be cautious about bathing in general, though they did approve of sponge baths. Manufacturers' educational pamphlets emphasized the necessity of regular local bathing to prevent odor and even cautiously favored tub baths but warned against the dangers of immersion in too hot or too cold water, or catching a chill from washing one's hair and not drying it quickly enough. In the early part of the century, many did not think that even sponge baths were a good idea because they were still likely to expose a person to cold water and air and chill her.

In this conflict between popular health beliefs and modern menstrual practices, sides were chosen along class lines. This class difference was most apparent in the stories that Ida Smithson and Roberta Cummings Brown told about their rural African American community, though it was evident in New England accounts as well. In addition to pointing out the meaningful distinction between those who threw away and those who washed their menstrual cloths, Ida distinguished between those who washed during menstruation and those who did not. She felt a certain amount of contempt for those who did not.

> I tell you, I knew some girls in school, that, they didn't change their towels, and when they got up, you could smell them, I can tell you that. I don't think they changed enough. I knew one or two in school. They didn't wash. This was another big thing. They were not told to wash your body, even though it was in their condition. But some people didn't wash during the *whole time,* and you know what that would do.

Ida herself had received no instruction from her mother and was extremely distressed at the messy, smelly result of wearing a single rag for too long during her first period, when she did not yet understand what a period was. She also did not receive any direct warnings against bathing, so she was perfectly content to adopt practices that prevented odor. In addition to bathing during menstruation, she discovered douching sometime after having her first baby, and she adopted that practice as well. "After menstruation. I would just, I was afraid that I may smell." A proliferation of bodily practices designed to prevent odor, in addition to practices such as striving for as much formal and informal education as possible for herself and her children, were part of Ida's trajectory into the American middle class.

During school, Ida and her friends did not simply ignore other girls' menstrual odor. While menstruation was not generally a topic of conversation, it could become a topic when classmates needed to be shamed into managing their bodies in an appropriately middle-class manner. "We would laugh about it, say, 'Gosh, don't let her get up!' you know. [Laughter.] So we were cruel!" Ida acknowledged that the teasing was cruel, but she also seemed to think that the girls who were teased brought it on themselves.

Those girls, who repeatedly refrained from washing during menstruation out of concern for their health, despite the cruel teasing, must have suffered tremendously from the conflict between their parents' health beliefs and the requirements of modern bodily practice that their classmates tried to impose on them. It probably was easiest to give in to classmates' "modern" ideas about menstrual hygiene. However, some girls maintained their parents' health beliefs and taught them to their daughters, who were likely Roberta Cummings Brown's classmates, twenty years later. The class distinction likely grew sharper by that time, and the teasing continued. Roberta heard from some of her classmates that it was dangerous to bathe during menstruation, and she figured that some of them followed that advice since "some of them were smelling pretty bad," but she thought it sounded ridiculous. By the time she told the story as a teacher to her middle-school students, it had become a fantastic tale of "the old days."

> I had heard that you weren't supposed to take a bath. Well, I knew that didn't make any sense. Because that was the time that you needed to take a bath. They said you weren't supposed to wash you hair and you weren't supposed to take a bath. And I had the sense to know that, gee, this is when I'm at my worst. I know that . . . What's the harm in it? So I did question that—why they said you weren't supposed to get wet. Well, that's the last thing that you should have followed. I always jokingly told the kids that when we would be talking about it. I said, "When I grew up, they told that you weren't supposed to get wet." I said, "You know how dangerous that would be." I said, "You'd be so scented!" So they always laughed.

Teasing and incredulous laughter distanced those who fully subscribed to middle-class ideals of bodily practice from those who were less "modern" and put health concerns about exposure to water ahead of concerns for cleanliness and odorlessness. It also served as an attempt to discipline those who failed to subscribe to the middle-class ideals. This distancing and disciplining, occurring both between social classes and between generations, was perhaps most strategically necessary among African Americans, though it was evident to some degree among other groups as well.[66] In American culture, African Americans were the most vulnerable to being accused of having failed to control their bodies or even accused of having bodies that could not be controlled. These accusations often served as one justification for denying African Americans access to the social, economic, and educational means to achieve middle-class status, and were therefore quite rationally feared and guarded against assiduously by those who aspired to that status.[67]

In the 1950s and 1960s, the use of deodorant powders on menstrual pads and vaginal sprays became popular, and African American interviewees were also especially likely to use these means of controlling menstrual odor. Deodorants had been advertised "for use with sanitary napkins" as early as the 1920s, and Kotex developed Quest, a powder deodorant specifically for menstrual pads, in the late 1930s, but it took a few decades for large numbers of women to begin using them. Amy Rivers sprinkled talcum powder on her sanitary napkins as a teenager. " I thought that everybody would be able to smell it, if I had an odor. And then everybody would know, right? [Laughter.]" Later, she used Quest and hygienic sprays, until she got an infection and her doctor told her it was from the deodorant products she was using. Faith Jones, born in the early 1950s to a high school principal and a teacher, objected to the smell of Kotex as much as to the smell of menstrual blood.

> Kotex would always to me have this, ew, you know, this odor that was there. Sometimes it wasn't bad, but it was, you could almost tell, OK, if someone's on their period. Not necessarily from the menstrual blood, because when that comes out, of course, it really doesn't have an odor. It's when it collects and stays. But the fabric sometimes would have a weird smell.

She used baby powder to hide odor in high school and later shopped around for more specialized products, choosing Summer's Eve over Vagisil "because it doesn't have much talc in it, so it works better. Talc is known not to be great. It can be irritating in the long run, or you can have the particles, and I wanted something that was safe, healthy." She also stopped using the sprays she had used when she was younger. "I really don't think they are that safe, to be constantly using something in that area of your body, you know, which could go internally if you aren't careful."

It was not only African American women who used powders and sprays; Christina Donvito, like some other white New Englanders,

> bought all the little products, and stuff, powders and soaps, and whatnots . . . I was always afraid I was carrying an odor . . . Hated that, just the thought of that . . . I think I became a little self-conscious about that. Just

wanted to be *sure* that if I was standing next to you, there wasn't a problem. You couldn't sense what was going on with me . . . I remember thinking that. Just wanted to be sure that I wasn't unpleasant to be standing near . . . It was like a security blanket [small laugh]. It was something extra I was doing. I can't tell you if it worked or it didn't work—probably didn't even need it! [Laughter.] But I did it.

Among interviewees, this was the only generation in which a large portion of women used deodorant powders and sprays to hide the odor of menstruation. Those reaching menarche in the 1950s and early 1960s came of age at a time of tremendous cultural consensus, centered around middle-class norms. They were the ones to most fully embrace the vision of the modern body articulated by advertisers and manufacturers, and the most thorough and deliberate in adopting the technologies and practices necessary to realizing this vision as it was developed early in the century. Their daughters, taking for granted the more sophisticated odor-preventing technology built into menstrual pads in the 1980s and 1990s, would question the safety and necessity of powders and sprays, as Faith did, and most would never use them.

As much as odor, stains on clothing was a cause of consternation and the target of preventative measures. Lillian Fang, born in the early 1920s to parents who struggled to support the family with their restaurant, remembered getting her period unexpectedly during a daylong meeting she attended in her capacity as a local government administrator. The memory remained clear even though the situation was not disastrous in the end. "Fortunately for me, I had a wine-colored dress on that day. And it was one of those kind of polyester dresses, so actually, I was able to wash it off during a break period, and it worked out OK. But I was so uncomfortable the whole day, worrying about what might happen." Like a number of interviewees, Liza O'Malley could remember, almost fifty years later, a menstrual accident that did not even happen to her, that she simply observed. "I remember very well being in grammar school, it must have been the seventh or eighth grade, and some girl had it on her dress and all the boys were laughing, and it was awful. And I made sure that would never happen to me. So I would never wear light clothes. I would always wear dark clothes." As with odor, boys participated directly through their teasing in reinforcing girls' determination to assiduously manage their bodies. Incidents such as this one confirmed women's belief that they would be humiliated if menstrual blood showed on their clothing.

All of the women interviewed for this book reported that their efforts to hide their menstrual blood were generally successful; accidents were so memorable partly because they were extremely infrequent. As Christina Donvito explained, this was the result of a great deal of effort.

> I always felt like I was walking with a mattress under me. Just to be sure. [Laughter.] It was never going to happen. Right? [More laughter.] . . . I did an excellent job on that. I might have gotten caught once or twice, where I really had to get home in a hurry. But after that—no, I did a good job. I made sure. Because it happens once, caught me twice, you'll never catch me again. You know what I mean? You just make sure. You just make sure you take care of yourself. That was just a part of your life that you had to deal with. It was ok, you know, I dealt with it. You can't fight it—it's there, you know?

Like Christina, many women found these practices rather onerous, but they did not question the need to hide menstrual blood. Roberta Cummings Brown taught her middle-school students that menstruation was "not anything taboo, it's nothing bad." Still, she explained, she was careful to let students know if they had stains on their pants and to help them get a change of clothes because "It's something that everybody does, but you don't want to do that . . . walking around with your clothes stained . . . It makes you uncomfortable." Roberta's students "were very grateful because it kept them from being embarrassed," since the requirement to hide menstrual blood, and the self-consciousness at having a stain on one's clothes, had not faded when the youngest of Roberta's students attended middle school in the 1990s.

With regard to stains, as with odor, African Americans were particularly sensitive to how their management of their bodies projected class status. Connie James, born to a lawyer and a teacher in an East Coast city in the early 1950s, remembered an encounter with a male friend. He was a medical student at the prestigious university where she was working and taking classes part time to finish her bachelor's degree in the 1970s. She was wearing white shorts, and he noticed a spot of blood, which he pointed out to her.

> He made me feel so bad. It was interesting. He was a black guy; he was giving me a lecture because I was in the middle of flunking a statistics class, because I wasn't going to class, and he said, "You're really acting like

a nigger right now." And what he really meant was, "You do not come from this class background that 'nigger' describes, and you're behaving as if you have to be from that class background, and here's one more piece of evidence of it, the fact that you're not taking care of your body properly." Basically, I bought his theory on that.

In both of their minds, educational achievement and respectable class status were tightly linked to care of the body, and it was reasonable for a man to remind a woman of her obligation to maintain the self-presentation and the level of education indicated by her parents' class status.

Beginning in the early decades of the twentieth century, and increasingly over the decades that followed, women sought out new ways to manage menstruation and looked to the mass market for technological solutions. Women's eager adoption of new, disposable menstrual technologies beginning in the 1920s accompanied and required a wide range of new practices and improved material circumstances. The new technology made periods a lot less uncomfortable but also introduced the managerial tasks of having to publicly purchase the technology, find a discreet place to store it, and dispose of it without clogging toilets or leaving visible or smelly remains in shared bathrooms. Women developed a host of practices to deal with these new problems. These practices became widely shared and taken for granted as integral to modern, middle-class habits of bodily care. They also began to develop new standards of appropriate discretion, whereby menstrual products could be visible within carefully specified locations and wrappers, but even so, the movement of menstrual technology into more public spaces caused many women a great deal of anxiety. New practices partially prompted, and were partially made possible by, changes in material circumstances, such as bathrooms with built-in cabinets and clothing that was affordable enough to be discarded if it was stained. The modern requirement to manage menstruation so that it was odorless and invisible became a standard to which almost everyone in the United States aspired by the second half of the twentieth century, and failing to meet that standard was explicitly recognized, and criticized, as a failure to attain a middle-class self-presentation.

By the last decades of the century, young women took "modern" menstrual management completely for granted. They developed more complex patterns of menstrual blood management to take advantage of the ways in which the new technologies could advance the long-standing goals of making menstruation less noticeable, less uncomfortable, and less of an inconvenience. This generation was the first to never have used belts to fasten their menstrual pads. Stick-on pads were introduced in 1970, and while the adhesive technology worked less than perfectly at first, young women took for granted that it was the only way they would fasten their pads. Most of the young women interviewed thought that using the stick-on pads the first time "was pretty self-explanatory." Anika Taylor explained that her mother "pretty much told me how to do it, but I didn't think it was that hard anyway, because you could pretty much look at it and see." Felicity Chang recalled, "I definitely knew what to do—I used a pad, and that was kind of it. It wasn't like a big deal. I was just a little bit annoyed that it had started."

Unlike the previous generation, while most young women first used whatever brand their mothers had in the house, many quickly became aware of the tremendous variety of products on the market and began shopping for their own. Maggie Yi described why it was so important to her to shop separately for her menstrual technology.

> My mom was really used to bigger pads, and once I discovered that not all pads were a foot thick, if something stated Ultra Thin, and another brand that was even bigger was marketed as Super Ultra Thin, I would clearly go for the Super Ultra, just hoping against hope that it would be really thin . . . I think that she generally tended to think that maybe it wasn't necessarily as absorbent. . . [but] I was convinced they were more absorbent. I was convinced at the time that thick was just not the way to go. So you're raised on these, like, people pouring beakers of blue liquid on the things on TV.

Many young women felt similarly, rejecting their mothers' Kotex in favor of brands that were making their products more absorbent and less bulky.

Most young women, like Maggie, were well aware of technological innovations in menstrual products, gathering information from television ads and friends. Felicity Chang appreciated the pace of innovations she had noted. "Over time as I've been using different products, I'm kind of struck by the core of engineers they have, really trying to improve these products. It's pretty interesting, I mean, there are lots of interesting products out there." She and her peers could name many brands and discuss the pros and cons of brands and designs, since they experimented widely before settling on the products they thought worked best for them. Ginnie Wang explained that her preferences moved "from Kotex to Sure and Natural, and then . . . it's been Always, you know, those super duper maxithins." Jennifer Kwan recalled the innovations she valued the most. "Whenever they came out with Wings, that was cool, when they came out with Dry Weave, that was cool, when it was a thinner material, but it was still just as absorbent, that was great." As far as brands, she described, "I actually love Playtex tampons, mainly just because they have slim fits, the little junior sizes, and also they have the smoother plastic applicator. I hate cardboard applicators . . . And I tried o.b. once, and I was like, 'I can't get it up! Am I supposed to shove it up more?' . . . In high school, I loved the Wings." Joan Randolph chose o.b. tampons out of environmentalist considerations because she did not like throwing away lots of applicators, and "Stayfree and their curved maxi kind of things with plastic on the sides."[68] Even men were aware of the available technology, from television ads. Adam Chiang could think of the o.b., Tampax, Freedom, and Kotex brands off the top of his head and explained, "I do remember that Wings was a new thing . . . in the last five years or so. Like, a cool new idea. And just how they always compare it to, like, a diaper, pour something in there and look how much it absorbs."

Most young women echoed Adam's observation that Wings, Always' trademarked term for tabs on the sides of pads that could be folded around the crotch of underpants and stuck to the outside, were an important innovation. In Amanda Chen's words, "Wings! The best thing ever invented!" Carlen Joyce Thomas, a businesswoman, tried them and thought immediately, "I wish I came up with this idea! [Laughter.]" She and others liked them "in terms of just saving your underwear—it was inevitable; [before Wings] I would lose a pair of underwear each month because [blood stains] wouldn't always come out." This generation expected their menstrual technology not only to prevent public embarrassment but also to prevent blood from leaving its mark at all. Because of this innovation, young women's brand loyalties shifted significantly from their mothers'. As Ginnie Wang put it, "I'm a firm believer in Always. Why would you do Kotex?" The brand that, to their mothers and grandmothers, had come to signify the whole category of products, seemed hopelessly passé and technologically primitive to this generation.

Young women's highly differentiated use of menstrual products was part of a more general pattern. This generation of women grew up with more meaningful variety in menstrual products than the generations before them, and they took advantage of this variety to manage their periods with more precision. Traci Anderson explained, "Somewhere along the line I realized that it was wasteful to use the same thickness all the time. It took a long time to realize that. It was a revelation. So then I would start buying multiple packages and trying to use them at different times."

While young women appreciated the technological improvements and choices of products, the planning could sometimes be burdensome. Heidi Xue, who worked in a small, male-dominated office, reflected,

> I always feel like in that week, it throws everything off, because I actually pay a lot of attention to it, in terms of, "Now, what do I have to equip myself with? . . . How much is my flow today?" . . . and it becomes sort of an ordeal in the morning. It's like, "Oh my God, am I meeting today?" or whatever. "Should I bring X amount of these, X amount of those? How many pockets do I have? Where I am going to lunch today? Will I be able to bring a purse?" That kind of thing . . . I guess I spend a lot of energy sort of trying to hide it. And trying to act like everything isn't any different from every day else.

The new technology was in many ways easier to conceal, and it did a better job of keeping menstrual blood hidden, but it also promoted young women's increased expectations of being able to do all of their normal activities, in their usual ways, while keeping menstruation concealed. In the end, this largely shifted the work of managing menstruation rather than reducing it, although it also made that work more effective.

A few young women briefly rejected disposable products on environmental grounds. Joan Randolph explained, "For a while, I was using reusable pads. They snap on. Because I was hanging out with my cousin at the time, who just had a baby, and she was using cloth diapers, and we decided we all should be using a renewable resource." Lisa Wood read about them with interest. "I think that making my own pads would be so cool, but I live in college. I don't have time to wash these things." In the end, Joan decided she could satisfy her environmental goals by using Instead cups, which were designed to cover the cervix like a contraceptive diaphragm and catch the menstrual blood; she reused them, contrary to the instructions on the box. Lisa did not find an acceptable substitute and put aside her environmental concerns for the time being. Both women discovered that modern menstrual management was woven into the fabric of their lives and could not be easily given up.

For young women in the 1980s and 1990s, modern menstrual management extended much further than Always super thin pads. They expected menstrual technology to live up to the promises advertised since the 1920s, and, in large part, it did. Jennifer Kwan related how as a teenager, "The discovery of Advil was great, and the discovery of tampons was great. Because the combination of the two made it that it felt like you weren't getting your period. There were no cramps, and there was no actual bleeding." Similarly, Ginnie Wang explained, "I want to feel like it's not restraining me. So anything that doesn't make me feel restrained on my period, whether it be Advil, or whether it be super thin maxipads. Those are the things that make me happy." Young women employed in many combinations the unprecedented range of technologies available to manage their periods, including birth control pills, over-the-counter Ibuprofen, thin pads with wings, tampons, and menstrual cups.

Young women were increasingly likely to be on the Pill, and as hormone dosages in the Pill dropped drastically and long-term medical data became available, their doctors felt confident prescribing it indefinitely. Many women born after 1960 began taking the Pill in their late teens and did not stop until they wanted to become pregnant. As Joan Randolph mused, when asked to compare what her periods were like with and without the Pill, "It's been so long since I've had the natural process going on, that I don't really remember." While Joan may not have considered the periods she had on the Pill to be "natural," she considered them her norm.

Most young interviewees went on the Pill for the purpose of birth control, but they were thrilled with the effects it had on their periods. As Jennifer Kwan related,

> I went on the Pill mainly for birth control reasons, but this was the nicest side effect ever, just to know when my period was going to show up. It would be every fourth Tuesday at 4:00. It just got so regular it was great! And it got much lighter, and fewer cramps, it made a huge difference, even just in lifestyle, in the way that I viewed having a period. Because even though it wasn't like I hated it, it wasn't something I looked forward to. But it would sneak up on me. And it just felt like, "Aha! Now I have the upper hand!" over my own body.

For Jennifer, the contrast when she went on the Pill was especially striking, since her periods had been extremely irregular and often quite painful. Even those who had no serious complaints about heaviness of flow or cramps found the predictability extremely useful. Vicki Liang, the daughter of academics who moved their family every few years around China and the United States during her childhood, explained,

> One of the pluses of being on the birth control pill was I'd know exactly when I was going to get my period, so I found this to be very useful when I was studying, when I had tests, so I could sort of like map out. I was like,

"OK, I know I'm going to get my period," and to prepare for it. So that was good. I actually liked that a lot, knowing. Because otherwise . . . it's this cat and mouse game. You have to get it right and carry tampons around with you at the right times.

This predictability reversed the process of calendar keeping for many young women. Instead of recording their periods as they got them and using the recorded information to guess when the next might come, they either looked at their Pill dials or wrote in their calendars for months in advance the days they knew their periods would be coming.[69]

For women who, like Joan, had thoroughly integrated the Pill into their experiences of menstruation and the way they thought about managing it, it could be a real shock to go off the Pill. Carlen Joyce Thomas described,

> I have gone off the Pill when I'm going to get pregnant, and . . . it's a much different experience. The cramps are much worse; it can last two days worth of being uncomfortable, maybe even longer; and the flow was to the point that you kind of had to make a mental note, because if you're not used to going to the bathroom constantly, and changing, and all that, you kind of had to stay on top of it.

For many young women, Pill use was thoroughly integrated into their management practices surrounding menstruation, and going off the Pill could upset a complex system of management techniques.

Some women, once they went on the Pill, discovered that they could use it to manipulate the lengths of their cycles. As Jennifer Kwan enthused,

> You can actually control your period, because you can take continuous packs. So that you can even skip it. That was like, amazing! Because I felt like earlier I was at the mercy of my period. It would just show up, whenever! And then it would interrupt my plans, or just ruin my clothes. And so now, being able to control it, and being able to control your body, is amazing . . . I just felt so powerful! I'm like, "I can conquer my period!" and I can control when it comes, I can control how heavy or light it is, and if I even want to see it again.

Many women were not as confident that using the Pill to manipulate their periods was a good thing to do and were not as successful in doing it as Jennifer. However, they still tried it occasionally. Joan Randolph explained,

I find myself fiddling to try and control when I get it, with my pills. Which is probably not too smart, but I don't do it too much. Like I've tried not taking the placebo pills [and starting a new pack right away], and that didn't work. I got it anyway. I think you have to do it more than a month . . . like they were saying in this article. Or I do, like, [my boyfriend's] coming next Wednesday, and I'm supposed to get my period Monday, so I stop taking my pills two days earlier than I would normally, so that I can get it over with faster. And it's not a big deal. He doesn't care. I don't care. But it would sort of be nice not to have to cope with this.

Brenda Xiao did not try to manipulate her cycles often, but "for my honeymoon, I was taking pills for like forty-five days, and I was just feeling like I was going to explode. So I didn't know if it was psychological or if something was really just wanting to come out."

Many women, like Brenda, had the feeling that it might be healthier not to manipulate their cycles too much, even though they appreciated the convenience of the light, regular, and easily manipulated cycles the Pill could produce. When Carlen Joyce Thomas switched to a Pill formula that her physician recommended as an updated, improved product, and her period got heavier, she "thought, 'Well, maybe this is better for my body, because it's more of a natural menstruation,' versus the other pills that would always change that." She asked her doctor about it and was surprised but glad to hear that the heaviness of Pill-generated periods was not an indication of their healthfulness.

American women born after 1960 were also eager and routine users of over-the-counter Ibuprofen to counter menstrual cramps, especially if their cramps were not already being relieved by the Pill. Ibuprofen was introduced as a prescription drug and sold as Motrin, beginning in the late 1960s. It was made available over the counter in 1984. Young women purchased it as Motrin, Advil, and Pamprin in various formulations that often also included substances designed to relieve menstrual bloating. It was the first medication that was both highly effective in relieving menstrual pain for most women and did not have immediate side effects. Young women used Ibuprofen regularly even when their parents thought that it was better to avoid medications, including those available over the counter. Ginnie Wang's mother was a longtime adherent to, and more recent practitioner of, Traditional Chinese Medicine, and thought that Ginnie should manage her health with Chinese herbal formulas rather than Western drugs. Ginnie was willing to go along with her mother's ideas when she was feeling tired or coming down with a cold, but when it came to menstrual cramps,

> after a while, it was not worth it. I only really use [Advil] for the first day, and I'm in such a bad mood and sometimes I feel yucky, that "Uh, who cares?" A painkiller here and there isn't going to do that much damage ... Generally, I just don't ever want to feel like I have to think about my period on the first day. So anything that will take my mind off. Even if it's psychosomatic, I take it anyway. I don't think I deal very well with ... a body that doesn't feel completely comfortable, or completely at ease that day ... I stay away from [Advil] beyond that because ... I'm a big believer in being natural.

Taking Ibuprofen for menstrual cramps was common enough among young women that young men "noticed when friends or colleagues are looking for pain reliever pills that their friends might have on them" and guessed that they had menstrual cramps. Adam Chiang wondered, "I know that there's PMS medication. I never really understood what that was. Like, how that was different from headache medicine, or just pain reliever. Like, what if I took Midol? What would that do to me? Is it just a general pain killer?"

—∞∞∞—

At the same time that young women adopted new technologies and management practices, taking modern menstrual management to new lengths, they expanded their expectations for support in their management efforts. First, they looked to advertising for useful information but castigated advertisers who misrepresented their experiences with menstruation. This generation was the first to grow up with ads for menstrual products on television, and they remembered them, gathered information from them, and commented upon them much more than their elders. Anika Taylor could describe particular ads for Kotex and Always that featured leakage tests of pads, and a tampon ad in which a girl going to college learns to use tampons from her older sister. "The tampon commercial I just laughed at, because I was like, 'It's *so* not that easy!' The pad commercial, I was like, 'Yeah, they really do work, I agree!'" Jennifer Kwan watched ads for new product features and new kinds of technology, both for herself and for her work as a sex educator, and she could describe some of the technological developments she had thought most interesting in recent years, including the Instead cup. She described herself "looking at developments, and really eager to try new things and new products. You know, like, 'applicator now smoother!' 'All right! I'll go for it!' But I'm really a sucker for advertisements."

Mothers of this generation sometimes noticed their children learning from these ads at much younger ages than they would have from print ads. Peggy Woo found herself discussing menstruation with her seven-year-old daughter who had seen an ad on television.

> My daughter just came up to me the other day, and said, "You know, Mom, they have these things, I think they're for your underwear, but they have these wings, and they look really cool. You should get some of those." [Laughter.] She says, "I'm not sure what they're for. I think it's for your underwear somehow." ... It's so cute. She thought I should have some. I said, "Yeah, it sounds great. Show them to me next time at the drugstore." She went into great detail, though, ... it's very cute.

Young interviewees were inclined to collect information about menstruation from advertising, but at the same time, they could be quite critical of the attitudes toward menstruation sometimes promoted in the ads and what they perceived to be unrealistic depictions of the experience of menstruation. Several young women shared Felicity Chang's objection. "I always remember thinking that it was really retarded the way they use that blue liquid. [Laughter.] ... That doesn't look like the consistency of blood. Nor it is the color, and there are no blood clots or anything, right? It's just this clean, sanitary, like, blue liquid."

Kim Chuang wanted to do away with the "feminine, flowery, pretty maxipad ad," because she did not think it reflected her experience of menstruation. "I feel yucky when I'm on it ... I am not excited about maxipads ... It doesn't work on me." Traci Anderson commented that "like any other kind of advertising, they suggest that if you use their product, it will be like walking along the beach, and it won't be any problem at all. They don't ever show women having cramps or anything else. It's just glorified—smoothes it over." These young women were the most successful of all the generations of women interviewed in approximating the vision of the efficient, leak-free, pain-free, modern menstruating body, but they were also the most insistent that the work that had to constantly be put into creating this kind of body be recognized and not be made to appear trivial.

In addition to expecting appropriate support and recognition from advertisers, young women began to expect a broader range of relatives, friends, and colleagues to acknowledge menstrual management and even to provide assistance. While many young women developed elaborate protocols to discreetly carry menstrual products to school and work, and then to the bathroom at the appropriate moment, a few interviewees thought that it was appropriate to reveal menstrual supplies even outside of drugstores and bathroom cabinets. Judy Han, born in the late 1960s to a Chinese immigrant farming family in California, explained, "I guess as I'm older, too, I'm not as shy about it, either. You know? I joke around once in a while that my pink, because it's always wrapped in pink, just sits in my purse. I should shove it to the bottom at least, or something." Kim Chuang was insistently open about her use of menstrual products at work, explaining, "When I go to the bathroom, I don't even hide [my tampon], I just hold it in my hand." One time, she stopped for a conversation on her way to the bathroom at the high-tech startup where she worked and was waving her tampon around for emphasis; she remembered this because one of the men in the conversation asked if it was candy, and she explained that it was a tampon. This matter-of-fact attitude was particularly striking to older women. Liza O'Malley told a story about generational difference.

> My youngest daughter was about six or seven, and my mother was babysitting, and my other, the daughter next to her was probably eighteen. There was something on the floor, and my mother said, "Oh, my God . . . What's that?" And Erin, the youngest one, said, "This is Janice's tampon." That was the first time my mother ever saw a tampon. She said she was mortified. But Erin was so nonchalant about it, the youngest one, that it didn't phase her. We talked, my mother and I, about how good that was, to not have that feeling that there was something wrong with it.

In keeping with both an expectation set up by their parents' generation's educational programs, and a newly articulated demand that their efforts at menstrual management be recognized, some young women decided that menstrual technology did not need to be hidden as it was being purchased or as it was being transported for use.

This broadening of contexts where young women were willing to mention menstruation was particularly striking to older women when men were included in the conversation. Mimi Cummings Wood contrasted her own reticence about talking to men outside her family about menstruation with her daughter's approach.

> I never would have approached a man during that time. And you know, times have changed dramatically. We went to Maryland a couple weeks ago to my niece's wedding, and so Lisa needs some pads, maxipads. So they had all gone out to a jazz club, and there was an older man that was driving

the car that she was in. It was lots of cousins and things. So I didn't go because it was the younger relatives. So when she got back to our hotel room, I noticed she had this pack of maxipads, and I said, "Where did they come from?" She says, "Oh, I asked Jim to stop so I could get some maxipads." I says, "You told him you needed to stop for some maxipads?" [Laughter.] It was the most ridiculous person [for her to ask].

Young men were not only asked to be sympathetic about menstruation but, like Lisa's friend, to help obtain menstrual technology. They might be reluctant and uncomfortable about buying menstrual products but could be coaxed into buying supplies for their girlfriends. They were asked to do so at a much younger (and likely more self-conscious) age than the previous generation of men. Kim Chuang tried to send a reluctant boyfriend out into the middle of a rainy night to buy her maxipads, and he suggested she just use his shirt. She thought that was "gross," and she was annoyed that he would not go buy them. She stormed into the rain to buy them herself, and he finally overcame his reluctance to run after her and accompany her. Alex Jones's college girlfriend asked him to go to the store late at night during his freshman year, and he agreed, despite his discomfort. "It was late, and there was hardly anybody in there. But when I got them, I still went to the counter and said, 'These are for my girlfriend.' I don't know why I had to tell the guy. [Laughter.]" He made sure to get very specific instructions about what to buy. "I was like, 'What do I get? I've been down that aisle before, but I never actually spend time there. [Laughter.]' . . . I said, 'Tell me exactly which one I need to get, so I won't be spending ten minutes standing there in that aisle.'"

Like Alex, many young men felt reasonably willing to help their girlfriends obtain menstrual technology but not very competent at it. As Ross Lyons pointed out, "Coming back with the wrong type of milk usually gets you some. You come back home with the wrong tampon, you're going to be in trouble." Christina Donvito's son, a lawyer, compared it to shopping for shampoo, where the options seemed overwhelmingly varied and he did not really understand the distinctions among products. Men perceived themselves as less competent consumers in general than their female partners and especially so when they had no experience with a particular kind of product and only a fuzzy understanding of how it worked.

Men were also expected to take an interest in their girlfriends' menstruation when their girlfriends had late periods. Because so many of this generation were sexually active before marriage, they talked to each other more often about the anxiety brought on by a late period and the relief of finally getting it. A few older women had mentioned this kind of thing to their female friends, but in this generation it newly became something for young men to discuss. Alex Jones remembered that at his high school in rural Virginia in the mid-1990s, "Guys who had girlfriends, you'd hear them comment on it. 'Oh, man, my girl said she was late this week!'" Adam Chiang had a similar experience in an urban California high school and college in the late 1980s and early 1990s, reporting that "People are always are in situations where guys are like, 'Oh, my God, she hasn't had her period' . . . It's certainly really scary." Young women shared their fears of pregnancy with their boyfriends, and young men took on a shared responsibility in worrying about it, monitoring their girlfriends' periods and looking to other young men for sympathy and guidance.

This generation of men may have been relatively willing to hear about menstruation from their sisters, girlfriends, and female friends, and to assist with it, because a number of them learned about it from their mothers, in addition to school programs. Deborah Leary sat her son down to discuss it with him, despite the fact that she left most of sex education up to her husband. "I explained about women and their monthly cycle, because my husband was too chicken to. [Laughter.] Just the basics, about what happens and why it happens . . . I think he was probably about twelve. Just so he'd know. Yeah, he didn't really ask a lot of questions." She knew her son was embarrassed by the conversation, but she felt it was important for him to know, as part of understanding sexuality. Not all women in her generation explained menstruation to their sons, but they still tended to care that they got some education about it. Laurie Wilson was comfortable having her sons know that she had her period, but she was glad that they had learned about it from a school program, so that she did not have to explain it. Scientific information could be accompanied by practical advice. Margaret Olsen gave some specific instructions to her sons, based on her own experiences with male peers in school. "I said, 'Don't ever be grabbing the girls' pocketbooks, and running around the playground with it, because she might have stuff in there that she needs, and it's very embarrassing for the girl.'" Sometimes supported by their mothers' educational efforts, young women expected young men to be sympathetic to their menstrual discomfort and the effort that went into managing menstruation, and sometimes corralled them into assisting with the labor of menstrual management as well.

During the first two-thirds of the twentieth century, American women and men collaborated with menstrual products manufacturers, advertisers and retailers to establish expectations about "modern" menstrual management and develop new technologies, practices, and standards of discretion to make it possible to live up to these expectations. New technologies and practices were supported by, and reinforced, the new modes of education and new health beliefs that were also part of modern menstrual management. By the last decades of the century, young women took for granted the new patterns of bodily management their mothers had worked to put in place, often at the cost of significant anxiety and long-term negotiations with manufacturers, retailers, and advertisers. They successfully managed their periods even more assiduously than their mothers and grandmothers, and at the same time demanded more support for their management efforts from colleagues and friends, including men.

As in their discussions of menstrual discomfort and PMS, they sometimes shocked their parents with their public recognition of menstruation. Seen in the context of the development of the modern period over the course of the century, however, this did not mark a moment of rebellion against the ways of the older generations. Rather, it was an extension of the logic of bodily modernity their elders had instigated. As far as possible, they made menstruation disappear. What they made more public was modern menstrual management, not menstruation itself. And what they demanded from their new audiences for talk about menstruation was generally either recognition for their management efforts, or direct assistance in their creation of modern bodies for themselves and their bosses, teachers, teammates, boyfriends, and anyone else who counted on them to be efficient and productive all month.

Like the development of modern education and health beliefs, the development of modern technology and practices is largely a story of a remarkably high level of participation and consensus on the part of women from different ethnic, regional, and economic backgrounds. While this level of agreement is crucial to recognize, it is not, however, the entire story. Disagreements and hesitations about tampons, among experts and laypeople, reveal the limits of consensus, and many interviewees' persistent self-consciousness about or unease with the Progressive, white, urban, well-educated origins of the modern period.

CHAPTER FIVE

Tampons

A Case Study in Controversy

Although women did not hesitate to use disposable sanitary napkins, many were not so confident that tampons were a good idea. Physicians were equally concerned during the ten years after tampons were introduced in 1936, and they studied and debated tampons' effects on health. Women and physicians alike were concerned about tampons' safety, efficacy, and sexual implications. Among interviewees, it was clear that these issues remained salient much longer for those outside the white, urban, well-educated middle class. Women who did not belong to this group at least recognized that they crossed social boundaries if they began using tampons, and some explained their decision not to use tampons in these terms. By the last decades of the century, young women interviewees almost all used tampons, their desire to instantiate modern bodies outweighing any persistent unease about tampons they might have acknowledged.

In the 1930s and 1940s, physicians debated the safety of tampons in major medical journals such as the *Journal of the American Medical Association* (*JAMA*), as well as local journals aimed at practitioners. A parallel debate took place in Great Britain, most evident in a long series of letters to the editors of the *British Medical Journal*. Physicians were sharply divided in their opinions about the safety of tampons. Previously, tampons had been medical devices, made of cotton, sometimes treated with medicine, that were "tamped" into bodily orifices to provide support or deliver medicine. Tampons were therefore not regarded as inherently dangerous, but using them to absorb menstrual blood meant that they were used significantly differently: first, they were used much more frequently, and second, they were controlled by the woman herself, rather than being prescribed by a doctor and managed by a nurse. Those who opposed tampons for menstrual management feared that frequent use would irritate the vagina and that women would introduce infection by not observing the same rules of hygiene as physicians and nurses. Detractors made this argument on the basis of principle, as well as some experiences with women who forgot to take out tampons at the end of a cycle and came to the doctor because they were concerned about an odor or discharge. Physicians expressed not only concern for women patients but also extreme distaste at having to remove long-forgotten tampons.

Physicians also worried about drainage of the menstrual blood. Through the early twentieth century, it was commonly believed that regular menstruation was necessary to rid the body of waste and that waste that was trapped inside the body would lead to illness. In the 1930s and 1940s, physicians who retained this concern translated it into a somewhat different language, seeing menstruation as equivalent to a wound. Physicians writing in the *Western Journal of Surgery* explained, "As an early surgical precept, we learned that whenever there is free serum, blood, or discharge from a wound or body cavity, free drainage is desired and much be encouraged."[1] Therefore, they argued, tampons were dangerous, potentially backing up the blood into the uterus, causing a great deal of harm.

Physicians who opposed tampon use generally argued from their own experience and intuitions, and from their understanding of basic medical principles. In contrast, physicians who were more favorably inclined toward tampon use, including several whose research was funded by tampon manufacturers, argued from experimental results rather than principles. They shared concerns about infection and blockage of flow but undertook research to find out whether or not tampons actually caused any of the problems they were speculated to cause. Not surprisingly, the research funded by manufacturers demonstrated little cause for concern; however, university-funded research gave the same results, which should have reassured medical journal readers that the chance of infection or blockage was not nearly as great as they feared.

An early study sponsored by the International Cellucotton Products Corporation (the subdivision of Kimberly-Clark which made Kotex and would soon market Fibs tampons) found some cause for concern, when it discovered that 18 out of 95 women experienced a gush of blood at some point on removing a tampon, indicating that the flow had been blocked rather than absorbed. The authors of the study concluded that this problem could be easily avoided by using smaller tampons and refraining from tampon use during the heaviest flow. As an additional caution, they recommended regular gynecological checkups for tampon users. This was the earliest of the experimental studies; its conclusion was cautious and acknowledged the common concerns from a physician's point of view. "Vaginal tamponage has been a medical procedure, and physicians have kept their patients under observation. Tampons are foreign bodies in the vagina. We do not know from experience how susceptible the vaginal mucosa may be to repeated irritation. When tampons are used regularly during the menstrual period, periodic examination of the vagina should be made to ascertain whether the mucosa remains normal."[2]

Later studies, done after tampons had been in common use for several years, displayed much more confidence that tampons were safe. A 1939 study concluded that not only did tampons appear safe, they seemed less likely than pads to introduce bacteria-laden material from the anus. Unlike in the 1938 study, the authors of this study did not think it was "best to exclude the subjective information obtained from questionnaires and from personal interviews" and noted that women who used tampons liked them a lot better than pads.[3] Findings in 1942 and 1943 studies confirmed these results, with greater numbers of women studied over longer periods of time.[4] Concerned doctors nevertheless continued to be cautious in their recommendations to their patients. In a 1942 survey of American and Canadian physicians conducted by two Kansas City physicians, almost 75 percent of the 211 who answered the survey were strongly opposed to tampon use, and, of the rest, about half approved of tampons only for limited use.[5] The same year, a new edition of a major gynecological textbook called tampons "a menace," and even the 1950 edition of the textbook opined that "Theoretically, they are harmful . . . Routine employment throughout menstruation cannot yet be accepted as innocuous."[6]

Even those who believed that tampon use was safe were not as persuaded that tampons were effective in managing menstrual flow. The research sponsored by International Cellucotton (Kimberly-Clark) showed that fewer than 10 percent of study participants could rely on tampons for complete protection during their periods, and 20 percent could not even rely on them during the last day or two of their periods.[7] These dismal success rates could be partially due to the types of tampons that were used. The study does not reveal the three brands that were tested, but one of them was almost certainly Kimberly-Clark's Fibs brand, which was about to appear on the market and which would be notorious for its inefficacy. Later studies, which used Tampax tampons exclusively, found that 90 to 95 percent of women studied could use tampons for their entire periods without having problems with leakage or blockage of the flow.[8] Presumably, the women who had success changed their tampons often; even with the higher absorbency of late-twentieth-century tampons, several interviewees found that they had to change them as often as every hour or two during their heaviest flow. It is possible that the efficacy of tampons seemingly improved not just because manufacturers improved their products or researchers chose to study the most efficacious products but also because women learned that tampons required different management practices than pads.

The concerns of physicians and the women who rejected tampons extended beyond concerns about safety and efficacy. In the first decades after tampons were introduced, physicians and lay people debated whether or not it was appropriate for young, virginal girls and women to use tampons. The concern was twofold: first, that the tampon would break the hymen, and second, that its use would be sexually stimulating and invite promiscuity or autoeroticism.

Doctors' concerns about sexuality were apparent in the above-mentioned survey of American and Canadian physicians, in which almost 75 percent of respondents disapproved of tampon use. The authors of the survey emphasized that many of those surveyed agreed that virgins should not use tampons; they did not mention the hymen but noted that "the difficulty of application was stressed."[9] Further, tampons were objectionable because in virgins, the tampon "brings about pelvic consciousness and undue handling, [and] may cause eroticism and masturbation."[10] These physicians assumed a model of women's sexuality in which sexual pleasure and desire was focused around the vagina and heterosexual intercourse; the fear was that the tampon somehow imitated heterosexual intercourse too closely. According to these concerned physicians, tampons had the potential to awaken sexual interest in women who were not supposed to be sexually active and to inspire women who did not have access to "the real thing" to discover unacceptable masturbatory substitutes.

Robert Latou Dickinson, a prominent and controversial sex researcher in the 1930s and 1940s, propounded a somewhat different model of women's sexuality. Dickinson took on the charges that tampons threatened the hymen and promoted sexual awareness. First, he compared the tampon to the sanitary napkin, turning the Kansas City physicians' analysis on its head: "The vaginal response in coitus belongs to the orifice and to the muscular girdle, as my swab tests show. The erotic stimulus of the stationary interior guard should be, therefore, momentary and negligible as compared with that of the moving pressures of the external pad, on areas provided with remarkably different ratios of sensory nerve endings." Expanding on the sexual dangers of the sanitary napkin, he pointed explicitly to its contact with the clitoris, noting that the napkin "is responsible for rhythmic play of pressure against surfaces uniquely alert to erotic feeling . . . The timing of this combination of warmth, friction and pressure fits the peak of congestion of the external genitals for the monthly cycle. Thus an unavoidable focus of attention on the region is emphasized for four days."[11] Dickinson did not, in this paper, challenge the assumption that sexual awareness in women, especially virgins, was a problem, but he did relocate sexual feeling away from the vagina and its assumed link to heterosexual intercourse. The traditional, acceptable method of menstrual management was shown to have been sexually stimulating young women all along, and the new method, which seemed so threatening, in fact removed unacceptable sexual stimulation, making it appropriate for middle-class girls.

Dickinson also addressed the issue of the hymen, detailing the measurements he had made of the opening allowed by the average virgin hymen, and showing graphically that tampons should fit easily into a virgin vagina and should not be difficult to insert or remove. He did not challenge the importance of maintaining an intact hymen but attempted to prove that tampons did not threaten this piece of virgin anatomy. Dickinson allowed the hymen to maintain its importance in conceptions of virginity but considered it a much heartier membrane, able to withstand significantly more poking and prodding, than the common wisdom of his time. He also advised that for those who found it difficult to insert a tampon, "a good educator is the douche."[12] Dickinson's conclusions, while perhaps not widely accepted, did reach a larger audience than just the physicians who subscribed to *JAMA*: *Consumer Reports* published a briefer version of Dickinson's *JAMA* paper, including his conclusions about the comparative sexual stimulation provided by pads and tampons and the finding that the average hymen could accommodate a tampon without breaking, while leaving out the detailed pictures of vaginal openings and tampon width. This publication marked a turning point in *Consumer Reports*' advice on tampons, which had previously warned against use by virgins.[13]

Tampax, meanwhile, from its first package inserts in 1936, insisted that only "in exceptional cases, very young girls may find it difficult to use Tampax properly. If there is doubt, consult your physician." By 1952, the package insert took on the issue directly, stating that "fully mature young women can use Tampax without impairing their virginity."[14] While Tampax educational materials insisted that virgins could use tampons without problem, materials produced by Kimberly-Clark, makers of Kotex and Fibs tampons, were more conservative. Their 1940 educational pamphlet *As One Girl to Another* told adolescent girls, "frankly, most authorities say young girls shouldn't use tampons without *first* consulting their doctors. The reason is this. In most girls, there's usually a membrane called the hymen which partly closes the entrance to the vagina—from which comes the menstrual flow. Therefore, Kotex sanitary napkins are more comfortable, and better suited to a young girl's needs, than tampons of any type." Of course, the

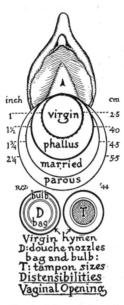

Fig. 4.—Caliber of distended hymen in virginal, married and parous woman, in relation to tampon and douche tube diameters.

Robert Latou Dickinson's illustration of vaginal and tampon width, published in the *Journal of the American Medical Association* in 1945

pamphlet was quick to point out that if the doctor approved, Kimberly-Clark's Fibs tampons were the best choice.[15] By 1946, in their pamphlet *Very Personally Yours*, less clear objection was expressed, though it was still recommended that a girl consult a doctor before using tampons; Kimberly-Clark clearly hesitated to recommend tampons to young customers. In a corporate meeting in 1952, an executive pointed out, "We have always been very conservative. I think we should still be. It is said, perhaps with some justification, that tampons should not be used by young girls, those just beginning to menstruate, 12–14 years old. It may stimulate some erotic tendencies, or at least their mothers might think so."[16]

It was not until the 1960s, when middle-class sexual mores were shifting in many ways,[17] that Kimberly-Clark moved to the position that tampons were smaller than the opening of the hymen and therefore perfectly fine for virgins to use; the question of eroticism was simply ignored. Tampax had good reason to push the envelope on tampon use by unmarried women, especially because these were their most eager customers, but Kimberly-Clark was content to recommend Kotex over tampons for that portion of the population until the bulk of physicians and sex educators had clearly given their endorsement to tampons.

By the 1960s, health columnists in magazines aimed at teenagers assumed that unmarried girls would be using tampons. By 1973, they were explicitly discussing the issue, declaring that first, virginity is a matter of abstaining from sexual intercourse, not having an intact hymen, and second, tampons should not break an intact hymen. For those who continued to be concerned about the status of the hymen, columnists were willing to provide reassurance, but they emphasized the newly mainstream idea in sex education that the hymen was not an accurate indicator of virginity or lack thereof. Of course, just because this had become the common wisdom among sex educators about tampons and hymens did not mean that this belief had spread to the rest of the population. In the late 1990s, they continued to answer letters from readers asking questions such as, "Everyone keeps saying that if I really am still a virgin, then I shouldn't be able to use tampons. But I have no problem with them. Does that mean my hymen is broken and that I am no longer a virgin—even though I've never had intercourse?" The reply emphasized that virginity is based totally on experience with heterosexual intercourse. By the late 1990s, once it had been thoroughly established in the mainstream literature that the hymen did not really matter, it could be admitted, unlike in earlier writings, that "using a tampon might stretch or tear your hymen," but it was OK, because "it will not change your virginity status."[18]

Tampons inspired as much controversy and ambivalence in the lives of interviewees as in the published debates about them. Interviewees, too, worried about safety, efficacy, and sexual implications. Often, their hesitations and self-consciousness about tampons reflected and revealed class, race, regional, and cultural differences that had been otherwise effaced in the almost universal embrace of other aspects of modern menstrual management.

First, many women who were not white, college-educated, American-born city dwellers found that they had to cross boundaries of race, class and region before tampon use even seemed like a possibility. Most of the Chinese American women interviewees born before 1950 did not recall knowing about tampons until they arrived in the United States. Many of the older women in all of the groups interviewed reported that they began using tampons when they went to college or heard about them once they married or had children. Roberta Cummings Brown described her experience going to college in the city after growing up in rural Virginia.

I went to college in the late fifties. And you had people from a lot of areas, so everybody's always pulling information about everything in those late night hours. Of facts of life, and things they had done, and things they wished they had done. And so the tampons, that would come up, how they were easier and safer. So that's how I, from then on, I did begin to use them.

Margaret Olsen, who grew up in a blue-collar household in the Boston suburbs, described how college friends, rather than her mother, introduced her to tampons. "I don't think she really was aware of them. It was kind of, pretty new, until college. Then different friends would say, 'I use them, I use them.' And I thought, 'Oh, if she can figure out how to use them, then I ought to be able to.' But I never really did until I was married."

While younger women were much more likely to have used tampons, especially before marriage, several of them found that they had to cross lines of class, race, or region before incorporating tampons into their menstrual management. Mimi Cummings Wood's daughter, Lisa, explained why she believed that during high school, most of her white friends used tampons but most of her African American friends did not. She thought it was

> not mostly about health. Mostly just about tradition. And I just would say the white families would be sort of more lenient, and sort of more, just more modern, if that makes sense. Instead of the black families . . . in most cases the grandmother's still in their period, because a lot of them have children younger. The grandmother's still in their period, mother's still in their period, they're on their period, and they all just use sanitary napkins, because that's what's in the house.

Lisa recalled how she started using tampons.

> I had this girlfriend who transferred in eighth grade, and she was from North Carolina, and she was a bad girl. I always remember she would talk about tampons . . . That's why I started using tampons, from Nicole Rivers . . . she was just like a big city girl. She went to concerts and all this stuff. And she had had sex before, and she used tampons . . . I thought she was really cool.

In Lisa's rural community, tampons were linked to the proliferating female sexuality associated with urban life; this was appealing to Lisa but appalling to many of the older women in her community.

Jennifer Kwan, who grew up in comfortable circumstances in urban California in the 1980s, attributed her tampon use to her immersion in white culture.

I feel like most Asian women, the more traditional they are, tend to use pads. They don't insert things like a tampon. They don't like even like looking at [whispered] "down there." Or inserting things. So maybe that's why my mom never used it herself, and she never told me about it. And it was just my friend, who was white—all of my friends back then were white—or [from the] dominant social culture.

It was necessary for many women that they go to college, have contact with urban American culture, have socially legitimated access to heterosexual intercourse, or accept the influence of white friends before tampons appeared as a viable technology to incorporate into their bodily practices.

Many interviewees were quite aware that tampons represented the epitome of "modern" bodily management, an ideal that originated in and reflected the values of the urban, educated, white middle class of the early twentieth century. In explaining why they did not use tampons, they often pointed to their own identities, and how they were different from that group. Liza O'Malley, born into a blue-collar Catholic family in a Boston suburb in the mid-1930s, knew about tampons as a teenager but did not use them. "I never knew anybody that did. I never knew anybody. None of my friends used them. Somewhere along the line I had in my head that that was for sophisticated people. More worldly people." Laura Hwang attributed her unwillingness to use tampons before she married to her embeddedness in Chinese culture. "Chinese people are very conservative. Before you are getting married, you won't touch things like, put anything in your body. Maybe that's the reason why I never, ever think about to use it [before marriage]."

Laura became an enthusiastic user of tampons once she immigrated to the United States and married. Liza tried them "because I was at the beach and I wanted to go swimming, and I also thought that meant that I was becoming sophisticated, becoming worldly." However, she felt that she failed miserably because she forgot to remove a tampon and a doctor discovered it during a gynecological exam. "When he pulled it out, the odor was so horrendous, and I was so mortified. I was so mortified." The second time she tried tampons, she had an allergic reaction that landed her in the hospital with a painful and embarrassing problem. After those incidents, she rejected tampon use and, to some degree, the "worldly" identity that tampons were useful in realizing. Not all women could or wanted to become "worldly," urban, college-educated, or culturally "white" enough to use tampons or create for themselves this particular aspect of "modern" bodily identity.

Family debates about tampons were in some ways similar to those of the medical and educational experts but not a simple reflection of them. It was often clear to mothers and daughters that in rejecting her mother's concerns about tampon use, a daughter was implicitly distancing herself from the aspects of her family's culture that were not compatible with white, educated, urban, native-born middle-class identity. Furthermore, many older women agreed with the doctors surveyed in 1942 that tampon use threatened virgin status. Women of all generations who chose to use tampons as teenagers reported conflicts with their mothers or at least a reluctance to let on that they were using tampons. Samantha Fried fought with her mother over her tampon use as a sixteen-year-old, in the late 1940s.

> I said I was going to the beach, and she said, "Well, you can't go," and I said, "Yes, I am going. I'm going to use tampons, and that'll be fine." And "Oh, you can't use tampons!" "Yes, I can use a tampon. It's perfectly fine." She started asking a whole bunch of questions. I said, "Calm down, no, that's not the case." She assumed that, she was concerned that I was off having sex somewhere, and that the only way you could use a tampon was if the hymen was broken—well, she didn't even know there was such a thing. But if you used a tampon, you had ceased to be a virgin. That was the big thing. And it wasn't being a virgin as such, it was if you had sex, you could become pregnant. That was the focus . . . Using a tampon meant to her that I was out there doing things that were going to cause big problems!

For Samantha's mother, as for many of the more conservative physicians and sex educators at the time, it was not clear whether tampons were a problem because they broke the hymen themselves, or because they indicated a hymen broken by other means. Either way, though, they were a threat to virgin status and a signal of proliferating sexuality. This debate would be repeated between girls and their mothers through the end of the century, long after health and education "experts" agreed that tampon use did not impact virginity or affect sexuality, although the debate seemed to be less fraught and less frequent by the later decades. For much of the century, to some women and some physicians, while tampons were clearly "modern," they threatened to undermine both traditional and modern middle-class sexual mores.

Among those interviewed for this book, Chinese immigrants and their daughters reported particularly prominent and ongoing conflict over tampon use. Most immigrants were not familiar with tampons until they arrived in the United States. Betty Li, daughter of a Chinese military officer and a homemaker who arrived in the late 1950s, had the impression that "when I first came, I'm not sure tampons were available. The pad was. Until later on." Even in the 1960s and 1970s, once tampons seemed commonplace and obvious to those born in the United States, many shared the experiences of Phoebe Yu, who explained,

> I know when I came to the United States, I went to the grocery store. I know that tampons were put together [on the shelves] with the Kotex, and I think it must be something. And that something, I wouldn't spend the money to buy a whole box, because I know I wouldn't use it, and then later on, I found in the office, in the restroom, or in public rest rooms, that you can buy both of them, and that's how I saw what it looks like.

By the 1980s on the Chinese mainland, tampons were advertised on television, but even for a young woman growing up in cosmopolitan Shanghai, they seemed foreign. Recent immigrant May Jue, born in the 1970s to an engineer and a high-tech administrator, described her surprise at seeing a tampon commercial as a young teen.

> In China we actually saw [an ad for] tampons, that thing that you stick inside. A woman was jumping up the board to swim, and she said, it's so comfortable that you can wear it to go to swim. I recall my sister, we all say it's a lie . . . Probably because we never experienced using that. Or the experience with having period each month is just so troublesome, we say, "It's impossible. It's a lie."

May continued to be skeptical about tampons after arriving in the United States. Women from China brought with them beliefs about health and virginity that made tampon use seem risky for anyone and a particularly bad idea for unmarried girls and women. This attitude was not universal; Laura Hwang, arriving from Taiwan as a young adult in the late 1960s, jumped to incorporate tampons into her bodily practices. "The first thing, I come to the United States, my friend teach me how to use tampon. And I say, 'Ooh, I like America, because of this!' [Laughter.]" However, Laura's enthusiasm was unusual, and even she waited until she married to use them. Betty Li, who also adopted regular tampon use, did not intend to use them before she married but got her period unexpectedly one day at work. "I was asking around, who has some supplies. And so [someone asked,] 'The tampon is ok?' I had never used it because I wasn't married then.

They said, 'It's ok, you know.' [Laughter.] It was a hilarious thing! I was in the bathroom; they were coaching me outside. [Lots of laughter.]"

Women who had health concerns explained them in various ways. Bonnie Kwan summed it up, speaking generally about Chinese health beliefs, as well as her individual concerns.

> We don't really like to use tampons because they think that they not be very clean and do a little bit damage on your body or something . . . I worry that germ, that I don't really want something in my body. And I think that I still have the feeling. Those are more in the blood so they have to be drained out of my body. And if you keep it inside, it's kind of a little bit toxic, no matter what.

Isabel Mao reflected, "To me I feel scary . . . I thought it's like a needle. Poke you. To me. I don't know. Right? You put something inside you. It's like a needle." Concerns about contamination, made prominent by Traditional Chinese Medicine beliefs about the necessity of draining menstrual blood and the vulnerability of the body during menstruation, may have been justified in terms of Western medicine as well, considering that in the 1970s, even relatively wealthy Taiwan was still a developing country with uneven levels of sanitation and manufacturing standards.

Women who had health concerns about tampons were dismayed when their American-born daughters wanted to use them and tried to convince them that tampons were unsafe. Jennifer Kwan's mother, Bonnie, explained carefully why her daughter should not have sex during menstruation or use tampons.

> She was saying, "Well, you know, it's probably not a good idea to have anything in your vagina (including a tampon) at the time you have your period, because the cervix is more open, to let the blood flow out, so germs and stuff can go in easier, so you probably don't want to be doing that" . . . [Also] she sees it as a plug of some sort, and that the menstrual blood is something, not so much waste, but it needs to flow out of your body, and so if you insert something inside it plugs it up and that's not healthy.

Sophia Lin was upset to discover tampons in her daughter's drawer because "What if it contaminated or something? I don't like. I heard something like that people got shock or something." The incidences of Toxic Shock Syndrome (TSS) caused by new synthetic material in Rely brand and other tampons in the late 1970s seemed to Chinese immigrants to be only expected, and the TSS warning labels on the backs of packages reinforced their concerns, based on Chinese health beliefs.

Some American-born daughters, in greater proportion than their peers who were not Chinese American, followed their mothers' way of thinking. Ginnie Wang felt, "If something is supposed to leave my body, I want it to leave my body as fast and carefree as possible. No pun intended on the brand. I don't like the idea of stopping something. I'm a big believer in flow." Heidi Xue did not use tampons until after college because her mother "had specifically mentioned toxic shock syndrome, over and over and over again. I don't know why, but it was her caution, and I never even tried because it seemed kind of scary. And of course, when you see a box of tampons, it's like in bright letters." Brenda Xiao remembered that she found out about tampons from girlfriends as a teenager "but just believed it was keeping the bad things in your body, like my mom had told me. And then, eventually, I just, I think just because of convenience, I didn't really care if it was keeping the bad things in my body." As she got older, her mother continued to warn her, telling her, "Don't use tampons because it interferes with fertility." Despite acknowledging her mother's concerns, Brenda continued to use them, feeling that the health risks did not outweigh the benefits of convenience.

Like Brenda, many Chinese American women born to immigrants chose to use tampons, and many of those were much less willing than Brenda to give credence to their mothers' health concerns. Angela Kong said that when she caught one of her daughters using tampons, she asked her daughter why. "Maybe because I never use that . . . I have certain reservations for using them. So, she say, 'Mother, it is very good, is very comfortable, is very easy. Nothing is in the way.' So I guess they use that." Isabel Mao was just as quickly dismissed by her daughter. "I try to tell her not to, see, because, to my knowledge, it's impossible, you put something inside. I said, 'Don't, don't, don't.' She laugh at me. She says, 'No, no, this is safe; this is safe.' So I guess that's safe. Because I think nowadays, the modern knowledge, they tell the schools, they teach the kids. [Laughter.]"

Isabel and others were concerned about their teenage daughters' health, but mothers also worried about preserving girls' virginity. Angela Kong felt, "When you are young, we never use that thing, to stick into your—what do you call it? Vagina? That's how I feel about it." Describing her daughter's use of tampons at age sixteen, she said, "She was always one of the black sheep in the family. She always would get ahead of everybody." Maggie Yi, born in the United States in the early 1970s, tried to explain what exactly her parents were thinking when they told her she could not use tampons.

Only people who are not virgins use tampons. And it wasn't phrased to me sort of as if like you will physically be unable to unless you've had sex . . . Rather sort of that like if you were using them, sort of that would be like a sign that [you were having sex]. It wasn't a very logical kind of thing, but I totally associated married women at that point in my life and tampons.

Tampon use, to Maggie's parents, was a clear signifier. Talking about a friend at school, Maggie "mentioned that someone had a tampon or lent someone a tampon and they would be like, 'Oh, so so-and-so is using tampons!' " When she went to a Chinese summer school, she was surprised and amused to find out that many Chinese parents had similar beliefs. Maggie felt that her parents used health concerns to mask what really worried them about tampons. "They mentioned Toxic Shock Syndrome, but that clearly wasn't their concern. I mean, that *clearly* wasn't their concern. And only my mom mentioned it. My dad never did."

While immigrant parents were concerned, in the end, most were resigned to their daughters' adopting bodily practices and altering their anatomy in ways that were acceptable in the late twentieth-century United States, even if they would have been unacceptable in Taiwan or Hong Kong earlier in the century. Florence Wu mused, "By the time I knew she was using it, she was already using it, so I said, 'Well, I kind of blew it.' I kept wondering, 'Gee, I wonder if that affects your virginity.' But then, I don't know, they go on trips with their boyfriends. It's a whole different story." Laura Hwang enjoyed American body practices herself, so despite her hesitation about tampons' sexual implications, she nevertheless approved of their use by her daughter.

> I have some kind of mind, you are not married, don't use tampons. But she use anyway, so, I just don't say nothing. She had used already. Besides, it's more comfortable, so I don't say anything . . . You don't feel nothing, and keep you dry and clean. I think that's her choice. I think it's good choice. Nowadays, they don't really care if you're a virgin or not.

These women decided that perhaps an intact hymen just doesn't matter anymore.[19]

They also sometimes wondered if their daughters based their decisions on better information than they had had growing up. Angela Kong thought her daughter had picked up tampon use "from her friends. I'm pretty sure. You see, these days, the young people, they talk about it. It's nothing to hide, or anything like that. They tell each other." While she approved of her daughter's openness, she may not have had great respect for her source of information. Others, though, like Isabel Mao, knew that their daughters had picked up "modern" information from school. Florence Wu asked in the course of her interview, "If you're a virgin, are you supposed to use that stuff? . . . But how deep is it inside? Because the tampon seems to go in pretty deep, doesn't it? . . . I thought, 'Why are all these girls—they are not married—why are they using it?' " She did not mean the questions rhetorically; she really was seeking information that she felt had not been available to her earlier. She again expressed her frustration with not having a good way to learn about menstruation herself as a teenager. "Here again, it's not having the knowledge. You're ignorant, you don't know, you have all these old wives' tales, all these scares, when you don't talk about it." Chinese American women were not always ready to abandon health concerns they learned in China, but they approved greatly of what they perceived to be the modern American way to giving young women lots of information. While most were not quite willing to withdraw their objections to tampons, both on the grounds of safety and on the grounds that they threatened virginity, they implied that life for their daughters in the United States might just be different enough from their own that their objections did not apply in the same way.

During the late 1960s and 1970s, as the sexual revolution flowered, tampon use became much more accepted among teenagers and young women. Among those interviewed for this book, many women born in the late 1930s through the 1950s gave tampons a try, even if they had hesitated to use them or had not even considered them as younger women. Many were under the impression that tampons were not marketed until the 1960s, an impression validated by a 1984 *Ladies' Home Journal* article, which listed decade-by-decade innovations in American society and technology, and placed Tampax in the 1960s.[20] In a sense, the article was correct because it was at that point that tampon use became widely accepted, even though tampons had actually already been advertised in the *Journal* for decades.

As more women tried tampons, they found that they not only had to cross boundaries of race, class, culture, and region to consider trying them in the first place but that they also had to adjust to the significantly different bodily practices tampons required. Those who liked tampons and those who did not noted the complications using tampons could involve. First was the difficulty of learning to use them. Peggy Woo recalled with humor the trials and tribulations of secretly learning to use tampons in a locked bathroom, as her mother asked what exactly she was doing in there for so long.

> I remember doing the Kotex, the belt, you know, that whole, horrible thing. Really uncomfortable, not working all that great, for a while, and then really wanting to try tampons. And my mother saying, 'No, you really don't need to do that; those things aren't really very good for you,' and that was sort of all she said. And so I found myself sneaking out to buy these little boxes of tampons [laughter]. And reading, opening the little instruction booklet, which of course we all just, throw it away, you know, don't even look anymore, but it's like, this whole manual, mini-manual in a box, of how the physique of your body works, and, just, the whole thing. Just total trial and error. And there was plenty of error! [Laughter.] It's pretty funny, now that I think about it. I just could not get those things in right for the longest time.

It was perhaps especially difficult for the younger of these women, who were more likely to be trying them as teenagers, and therefore more likely to have to be surreptitious as they experimented with tampons.

It could take some experience for women who waited until they were older to use them, too, though, before tampon use really became routine. Deborah Leary did not use them in high school, in the early 1970s, since her mother did not have them around, her friends did not use them, and it did not occur to her to go buy them for herself as Peggy did. "As a teenager, [menstruation] was more of a nuisance than anything. I hated the pads . . . Not too many girls used tampons at that time. I just remember it being messy. And then once I hit college, I started using tampons and that was easier." It was not easier right away, though. "It was very uncomfortable, and the first time I'm sure I had it in wrong, totally wrong, because it hurt. But I got better. I didn't give up. [Laughter.]" Amy Rivers remembered rejecting Kotex and starting to use tampons "as soon as I got there [to college.] 'I'm not using these things anymore!' And I do remember using up almost half a pack trying to insert it. Trying to get it right." Laura Hwang heard about them from a friend who was also an immigrant from Taiwan and persisted in learning to use them despite the fact that

> the first time, I did not put it in the right place, and I thought, 'Oh, my gosh, it hurts!' And I could not take it out, I had to soak myself in the water to take it out. [Laughter.] And then I learned it. After I learned it, I loved

it . . . If somebody says it's good, there must be some reason—you have to be open-minded. You know, if it hurts, that means you don't do right. You have to give yourself a chance, right?

Peggy, Deborah, Amy, and Laura all persisted in learning how to use tampons, despite sometimes substantial initial difficulties, and used them throughout their menstruating years.

Some other women, however, were not so enamored of them. The second problem with tampons was that even after learning to put them in, it took a lot of experience to learn how to judge whether a tampon was full of blood at a given time, and not everyone found their bodies to be especially predictable. Barbara Ricci summed up the problems she had with them. "It can be uncomfortable if you don't actually have a lot of flow. Because they're kind of dry when you take them out, and put them in. And then if you have a heavy flow, they're often not enough anyway. And it's just harder to tell what's happening—you can't see it." Florence Wu tried Rely tampons, the particularly absorbent brand linked to Toxic Shock Syndrome, when she was in her forties, and found the new material used in it exaggerated these problems for her. "I think the worst thing, the one time that I completely said, 'forget about it,' there was a particular tampon that bloated up really big. I can't remember—they took it off the market. I used that, and it was so painful to get out, and I said, 'Never again. I never want to use a tampon again.'" Margaret Olsen did not find tampons uncomfortable, but, she explained, "I was afraid it was going to leak out. I just couldn't believe that this was going to really work . . . I was constantly going in and checking, and checking. And then going out, and thinking, 'Is it working? Is it working?' And then I thought, 'The heck with this!'" Tampons were significantly harder to monitor than Kotex and, for those who had not taken the time and effort to develop the requisite managerial skills, could cause their own kind of discomfort and risk of embarrassment.

A third new difficulty that tampons introduced was the possibility of a woman forgetting one at the end of her period and leaving it in, or believing she had taken one out and inserting a second. Barbara Ricci remembered that she once had a problem with

> a tampon. In college. Kind of a funny, strange situation—I went out to a party or something, drank a little too much, thought I had pushed one in too far, couldn't get it out, and then I guess by the time I got home I was ok. I was worried about it, I called my mom, long distance, said, 'What do you do if . . . ' and she said, 'Well, you just *have* to get it out. You just reach

up in there.' And my sister-in-law happened to be there, too, who was a nurse, and tried to tell me, 'You'll be able to get it; just reach up and find the string. Call back if it's really a problem, and you can go to a doctor.' I don't know if it's because I was a little inebriated or what, during part of the evening, but I must have at some point—basically, I never found it! [Hearty laughter.] I must have at some point gotten it out but totally forgot. This was a totally strange thing. So, it was something I never forgot. I still wonder where it went. But I had three kids, I know it's not still in there. [Laughter.]

While that episode was not enough to scare Barbara away from tampons, the difficulty of gauging which absorbency to use at a given time, and then the deaths from Toxic Shock Syndrome linked to Rely tampons in the 1980s, persuaded her that the benefits of tampon use did not outweigh the potential problems they created. More women coming of age in the 1960s and 1970s felt impelled to try tampons than did women from earlier generations, but they were also more likely to provide a thorough critique of tampons based on their experiences, and many made the choice to go back to Kotex.

Although a number of interviewees who grew up in the United States in the 1960s and 1970s decided not to use tampons themselves, unlike immigrant mothers, they generally did not object to their daughters using them, as long as they took the box label's recommended precautions for avoiding Toxic Shock Syndrome. They had made their own decisions about tampon use less out of health concerns than out of a belief that tampons were more trouble than they were worth, and if their daughters found them convenient, they had their mothers' blessings. When Margaret Olsen's daughters began using tampons as teenagers, she recalled, "I thought, 'That's great. I'm glad they got it.' [Laughter.] And in fact, I can remember one of them saying, 'Uck, I can't believe that you'd ever wear those pads, and have all junk just sitting there!' . . . And that's kind of always been their perspective. 'What a disgusting thing, ugh.' And it really is! [Laughter.]" Barbara Ricci saw their usefulness for particular situations. "I do think they're wonderful for people who exercise or swim. Or dance, or whatever, where you don't want a Kotex to be seen." Though Margaret and Barbara rejected tampon use themselves, they did not reject it in principle and encouraged their teenage daughters to use them.

This meant that their daughters, born in the last decades of the century, sometimes learned how to use tampons from their mothers. As a young teenager, Serena Joyce Ambrose saw her mother and older sister using tampons and wanted to use them too. She asked her mother, and her mother showed her the directions, and explained what to do. "It was awkward at first, but ever since then I've preferred them over the pads." Joan Randolph's mother tried to teach Joan to use tampons when "one time I got my period and I had nothing, nothing in the house." It was evening, and they lived in a rural area, so getting to the store seemed difficult. Joan tried it but could not get the tampon in comfortably, so she gave up and used toilet paper instead. She guessed later that it had been difficult because her mother used super-size tampons; she finally did start using tampons when she was a little older and purchased the junior size, which she thought easier to use as a beginner. Joan's experience was a striking shift from that of most women of her mother's generation, who usually hid tampon use from their mothers, assuming it would be a potential source of conflict.

Young interviewees were likely to have begun using tampons during junior high or high school, not waiting until college or marriage as most of their mothers had. Taking for granted modern menstrual management, they had high expectations of the related technology. While they appreciated the recent innovations in pads, some still had complaints. Several explained their choice by describing sanitary napkins in terms that would have sounded familiar to Mary Hanson at the beginning of the century, when she described cloth napkins. Kim Chuang thought that learning to use o.b. tampons from her friend "was very a empowering feeling . . . It didn't feel like you were on your period. I loved it." In contrast, she described, "When you're wearing a pad, it feels like you're carrying this weight around. Especially because I used the really big ones—it's like you're wearing a diaper." April Chandler bought herself tampons and learned how to use them by reading the directions on the box, "because my mom just used the pads, and I didn't like those because they felt like diapers to me. [Laughter.]" Mary Hanson would have laughed at these young women who thought that late-twentieth-century, relatively thin, adhesive Kotex was like a diaper, but in fact, they did have similarities to contemporary diapers in the same way that the cloth pads Mary made for herself resembled the diapers used for babies for much of the century. Young women did not like the awkward, infantilizing feeling of menstrual pads any better than did Mary, and they also sought alternatives—for this generation, primarily tampons.

Young women also particularly complained about the physical sensations involved in using pads and chose tampons to avoid them. Echoing Margaret Olsen's daughters, Lisa Wood put it, "[Do] you find it more appealing to have this blood laying against you [laughter] or just contained in a little, small thing?" Kim Chuang complained, "When I'm wearing a pad, I'm constantly reminded. When

I'm on my period, I feel it coming out. The most disgusting feeling in the world. You just feel it dripping out." Young women not only expected to be able to avoid the sight and smell of blood, and the awkward feeling of wearing a thick pad, they expected to be able to avoid the feeling of contact with menstrual blood.

Besides issues of sensation and comfort, many began using tampons because they wanted to swim or play sports during their periods. By this generation, few women thought they should refrain from activities during menstruation out of health concerns, though many felt uneasy about using tampons at first or had mothers who had concerns about virginity. Brenda Xiao explained why she finally overruled her mother's objections and started using tampons.

> She'd always make me wear a pantyliner when I was swimming, and then one time, it came up. [Laughter] In the pool. I've had terrible menstrual experiences, now that I think about it. So it's floating on the surface of the pool, and we were swimming with a bunch of friends, and the guys were like, "Ew, ew, whose is that? Whose is that?" So I totally shut up. I'm like, "Not mine!" And I think after that I started using [tampons] for swimming. And then eventually it just moved into [using them] for everything.

For young women who thought they should be able to do all their normal activities during their periods, correct technology was crucial to avoiding embarrassment.

Of the young interviewees, Chinese Americans were the most likely to avoid tampon use. However, there were a few young women from all groups who chose not to use them. Young white women often shared Chinese Americans' concerns about TSS. Traci Anderson avoided tampons out of concern about TSS for a number of years and then simply decided that pads worked well enough that she did not need tampons. Rose Mitchell avoided tampons for years because "I have a tendency to take rare possibilities and think it's going to happen to me. So the main reason why I was afraid to use tampons was I'd heard about TSS. And was convinced that if I tried it, I'd get it, and I'd die. [Laughter.]"

Young women found that it was not necessarily easy to stick with a decision to avoid tampons. After high school, most young interviewees found themselves in communities where it seemed that everyone was using tampons. Their mothers learned in college that tampons were a reasonable possibility; this generation felt pressured to begin using them during college in order to be as discreet, efficient, and capable as their peers. Maggie Yi described, "I never actually had seen an unwrapped tampon until college. My roommates all used them, and they thought it was hysterical that I hadn't. I remember that." Even in high school,

Rose Mitchell was extremely self-conscious about having her period, partly because "most of the girls that I knew at school [used tampons]—and I think that was part of the reason I was embarrassed about it, I didn't use tampons, I always used pads, and they were bulky, and obvious, and not as neat. So that's why I think I had kind of a tendency to try to hide it." When she got to college, "in the dorms, especially being on a floor with all women, you couldn't be private anymore. Everybody knew, pretty much, what products you used . . . So it was time for me to grow up. And because I was convinced that nobody else used [pads] but me." Rose was pleased with tampons once she got used to using them and felt less embarrassed about having her period once she felt less self-conscious about the technology she used to manage it.

Young women could bring a lot of friendly pressure to adopt particular technologies and practices to bear on their friends, since they were much more likely to be talking to each other about their menstrual practices than were women of earlier generations. Amanda Chen described,

I have a friend from college who still, as of last summer, had not [used tampons]. All of her female friends were encouraging her to try it . . . All this coaching from all of her friends, and, "Oh, you've really got to do *this* or *that*" . . . The first couple times she was really unsuccessful, and I'm sure it was because she was nervous. Or she tried and would keep it in for fifteen minutes and be all uncomfortable and take it out. And so I remember getting an email at work, "I did it!" [Laughter.] I laughed so hard. I called her, I'm like, "Oh my God!" It was such an accomplishment for her.

Friends both enforced the necessity of adopting modern menstrual management practices and supported each other in efforts to do so.

Managing their menstrual flow with tampons, for this generation, also meant managing the risk of TSS. As noted above, some young women were very concerned about TSS and avoided tampons completely because they were afraid of getting it. The syndrome appeared in the news sporadically over the 1980s, first when a pattern of TSS linked to tampon use in women was discovered by epidemiologists, and then when TSS victims brought lawsuits against tampon manufacturers, and manufacturers and watchdog groups wrangled over whether and what kind of warning labels should appear on boxes.[21] In 1980, when TSS linked to tampons was first announced, women's tampon use dropped off sharply.[22] Tim Dai remembered that while he was in junior high school, during the mid-1980s, it seemed to be in the popular press for a little while. And I just remember learning about that and being kind of disturbed by that and being concerned for people and not really knowing what its general implications were. I think I was also under the strong impression at that time that my sister and my mother were both using pads instead of tampons and so I remember being kind of aware of that.

Many women waited to resume tampons use until it became clear that the epidemic of TSS seemed to be the result of a new, more absorbent material, present in particularly large amounts in Proctor and Gamble's super-absorbent Rely tampons, and Rely was taken off the market. Still, TSS and its link to tampons were not well understood, and while health specialists could recommend practices that they believed would significantly lower the risk of TSS, they could not give any guarantees.

Despite concerns, many young women thought about the risk as something that could be reasonably managed if they took appropriate precautions and carefully planned their use of tampons. They were more aware than their mothers of TSS and had considered it when they planned how to manage their periods. Carlen Joyce Thomas "had a little system. At night, I always wore a pad. During the day, I would pretty much wear a tampon." Amanda Chen also

> kind of went back and forth. I don't wear tampons at night . . . I worry a little bit sometimes now, because I feel like if I'm using a tampon that's too strong, for what little flow I have now [on the Pill], that it's worse . . . The minute I think it's more doable, just changing more often, I'll move to pantyliners . . . When I was on my regular flow, not on the Pill, I liked the Kotex, the plastic applicator ones. They're thicker, but somebody told me in high school that they're not that condensed cotton. The absorbent thing is not as condensed, which makes it somehow, you're less likely to get TSS.

Amanda and others were willing to spend significant energy to plan their use of menstrual products carefully, in order to have the convenience of tampons while minimizing the risks associated with them.

Young women were very conscious of TSS when they felt they had made mistakes in their menstrual management. Kim Chuang once forgot to take a tampon out at the end of her period and went to her gynecologist because she was concerned about a smelly discharge from her vagina. When he pulled out the tampon, she said, "I was so embarrassed! I was like, 'I can't believe that was in there for a month. I didn't feel anything for one month.' I was totally tripping out, I was like, 'I could have died from Toxic Shock!' He's like, 'No, no—that's really rare, don't worry about it.'" Her memory of embarrassment came first, but her concerns about TSS were not far behind. Young women knew they took serious, if small, risks in using tampons. They managed risk as part of managing menstruation and urged this risk on their peers as well, because they believed wholeheartedly in the benefits of producing discreet, efficient, productive, capable bodies all month.

Young interviewees' almost universal acceptance of tampons, despite recognized risks, came at the end of a process of about fifty years of experts and lay women alike debating safety, efficacy, and sexual implications. These debates, public and private, revealed the origins of the modern period in the Progressive era's white, urban, well-educated middle class. Many interviewees felt a sense of unease, or at least self-consciousness, about using tampons based on how their own identities differed from these social characteristics. The nature of concerns about tampons, and the social boundaries they revealed, were particularly evident in conflicts between mothers and daughters over tampon use. By the end of the century, however, even a persistent self-consciousness about social identity and tampon use did not prevent young women from contentedly using them as a key technology in modern menstrual management.

Conclusion

Over the course of the twentieth century, Americans radically revised their ideas and practices surrounding menstruation, to bring them in accord with a desire to become "modern" and attain middle-class standing. American women adopted Progressive ideals of efficiency, education, and good management, and applied them to menstrual management. Both women and men idealized a body that could work and play at full efficiency all month, and a way of handling menstruation that would force it into the background of self-presentation and bodily sensation as much as possible. In configuring the modern period, they collaborated with a variety of experts, many of whose roles had also emerged out of the Progressive impulses and large-scale social changes of the late nineteenth and early twentieth centuries. Sex educators, physical educators, industrial hygienists, advertisers, and menstrual product manufacturers worked to give shape to new desires and ideals, often in self-interested ways, but also with an eye to what women would find persuasive and worth adopting.

Women and experts alike emphatically rejected ways of managing menstruation they regarded as old-fashioned and that indeed generally had roots reaching back centuries, even millennia. Like many generations of their forbears, interviewees born early in the twentieth century typically learned about menstruation haphazardly, used and reused cloth rags to manage bleeding, and hesitated to bathe or swim during their periods for fear of negative health consequences. They explained how these pre-"modern" ways of handling menstruation left them feeling ashamed, confused, physically and emotionally uncomfortable, and vulnerable to public embarrassment. They felt that "modern" ways alleviated much of their discomfort and much better supported the self-presentation and range of activities they and others were coming to expect of them all month.

Changes in three areas—menstrual education, health beliefs surrounding menstruation, and menstrual technologies and practices—constituted the creation of the modern period. First, sex educators wrote books and pamphlets designed to be given directly to girls, rather than through their parents, and began to disseminate them through schools and libraries. Kimberly-Clark and other manufacturers became the most important distributors of menstrual education materials, reaching millions of women by the end of the century, and urging that girls be informed before menarche. Women began to look to published sources of information containing scientific explanations of the menstrual cycle and, as interviewees explained, saw them as positive, helpful sources, much more thorough and reassuring than most of their parents had been. They began to give pamphlets to their daughters at or before menarche, using them to facilitate a mother-daughter interaction they otherwise found awkward and difficult. Interviewees born in the 1940s and 1950s were most enthusiastic about modern menstrual education and were most likely to seek out information themselves as teens. When they had children themselves, they tended to turn to modern, scientific explanations to talk about menstruation with them, and they sometimes allowed their children to witness and help with menstrual management in the bathroom as part of sharing information with them before menarche. The sex education texts themselves and the novel educational venues in which discussion of menstruation was made newly acceptable were important in helping women feel that they could manage menstruation more effectively, efficiently, and matter of factly.

In addition to modern menstrual education, women and new varieties of "experts" adopted a new set of beliefs about health and menstruation, gradually abandoning millennia of concern about the necessity of regular menstruation to general and reproductive health. Starting with the "new women" of the late nineteenth century, they began to challenge the idea that women needed to take special precautions during their periods to maintain their general and reproductive health. Physical educators and industrial hygienists established new norms for all-month play and work in colleges, factories, offices, and department stores. By midcentury, the idea that emotional shock during menstruation was dangerous had been long forgotten, "modern" women bathed without worry during their periods, and many of those interviewed for this book believed that swimming during menstruation was an issue only because it was difficult to manage the blood in that situation, not realizing that their foremothers had serious health concerns about the practice. While they did not necessarily rush to break taboos against sex during menstruation, they agreed with twentieth-century sex researchers that it was primarily a question of aesthetics and convenience, not a health or moral issue. They no longer thought about menstrual health in terms of needing to preserve regular cycles to protect their health but in terms of relieving the discomforts caused by menstruation. When they turned to doctors for help with menstrual pain, they were often quite frustrated, finding that doctors never seemed to offer true relief except accidentally, as a side effect of birth control pills. They believed in the modern period and expected a whole range of experts, including their doctors, to help them achieve it.

The third area of change, in addition to menstrual education and health beliefs, was in menstrual technologies and practices for managing bleeding. Kimberly-Clark introduced Kotex as a highly advertised, mass-market product in 1921, rapidly popularizing standardized, mass-produced disposable sanitary napkins. Advertisers created some of the most deliberate and explicit images of what it meant to be a modern woman, linked them tightly to the new technology, and distributed them extremely widely. Women and retailers created new practices around new menstrual technologies, carving out specific spaces in drug stores and bathrooms where menstrual products could acceptably be seen in public. These carefully specified and organized public displays in turn supported the enhanced discretion of modern bodily management in other circumstances. Women used new menstrual technologies to meet the demands of modern bodily presentation, doing their best to eliminate the stains, smells, and discomfort they associated with menstruation, to make menstruation as undetectable as possible to everyone, including themselves.

Changes in menstrual education, ideas about menstrual health, and menstrual technologies and practices very much reinforced each other. Sometimes the reinforcement was direct. Using tampons made it a lot more comfortable to play sports all month, for example, and menstrual education pamphlets included information about modern menstrual products and health practices as well as a scientific explanation of the menstrual cycle. But even where such direct connections were not apparent, the various new experts, technologies, and practices all aimed at the same ideal of a modern body, which was well managed, efficient, productive, and predictable.

This coherent vision of modern menstrual management, well supported by education programs, widely circulated health beliefs and expectations, and standardized technologies and practices, was taken for granted by young interviewees in the last decades of the century. They expected to manage menstruation easily according to "modern" standards and were highly critical of experts and technologies that failed them. They extrapolated their parents' and grandparents' patterns in ways that sometimes shocked the older generations, often requiring female and male peers alike to provide sympathy and assistance for their menstrual management efforts. While their elders were sometimes taken aback by their actions, interpreted in historical context, the youngest interviewees were simply extending the reach and effectiveness of modern menstrual management by carving out a few more specific ways in which menstruation could be made public in service of modern management practices.

In historical time, this emergence of the modern period happened amazingly quickly. Over the course of less than a century, Americans abandoned millennia of health beliefs and practices around menstruation and developed a coherent model of modern menstrual management supported by a bevy of experts, institutions, and technologies. In the context of an individual's life, however, these changes could seem quite gradual. Many older interviewees adopted modern ideas and practices only piecemeal, or gradually over the course of their lives, and often experienced the change as a struggle. Their stories about tampons revealed their self-consciousness about the Progressive roots of the modern body, in urban, white, highly educated culture, and their frequent unease with the ways in which they felt this contradicted their personal histories and identities.

Interviewees generally felt satisfied with the modern ways they had adopted to manage menstruation. Older interviewees strongly preferred it to what they remembered doing before, which was the alternative that was most immediately obvious to them. Some young interviewees tried some alternatives out of the mainstream, for example, washable menstrual pads designed to be environmentally friendly but found that these alternatives could not support the patterns of work and play they took for granted. Most found that modern menstrual experts and technologies generally delivered on their promises, and after the awkward first months or year after menarche, they felt competent and relaxed in their menstrual management practices most of the time.

Scholars have focused, understandably, on more fraught areas of modern experience and bodily practices and wondered why so many women have been willing to submit to constant dieting, cosmetic surgery, or highly medicalized and dehumanized hospital births as a matter of course.[1] This book's analysis of the modern period suggests one reason why Americans may have developed so much faith in techniques of modern bodily management; in some areas, it lived up to expectation. According to those interviewed for this book, managing menstruation in a modern way relieved the shame, anxiety, and discomfort of older methods, and usually allowed women to pursue their work and play as they and others had come to expect. A new mode of bodily management that enabled these activities perhaps felt especially liberatory to women, who otherwise often experienced menstruation as something that kept them from competing and participating effectively in schools and workplaces with male peers.

It also served as one basis for Americans across widely disparate standards of income and education to feel that they had achieved middle-class status. Women and their husbands benefited from women's modern, middle-class self-presentation and the cultural and workplace status it helped them to achieve. The demands of modern menstrual management, in the end, were not necessarily onerous compared to premodern management techniques, and were much more consistently rewarding. While they created a new kind and volume of managerial work for women, they did not seem to come with the kind of constant striving, perpetual anxiety, and sense of frequent failure found in, for example, dieting practices.

Given the success Americans felt they had with modern menstrual management, it does not appear that we will be leaving the modern period behind anytime soon. Its latest instantiation, an early twenty-first-century twist on modern menstrual management, comes in the form of extended-cycle birth control pills. In 2003, Barr Pharmaceuticals introduced Seasonale, an oral contraceptive designed to be taken for eighty-four days in a row, followed by seven days of inactive pills, allowing for one bleeding approximately every three months. This idea was certainly not new. Doctors had been prescribing birth control pills, off-label, with similar instructions for a long time, and many women, including some of the women interviewed for this book, had taken their pills similarly, with and without the blessing of their doctors. However, Seasonale was the first product to receive Federal Drug Administration (FDA) approval for this use and the first to be advertised in the medical and popular presses as a birth control pill designed to allow only four periods a year.

While Barr had to demonstrate that Seasonale was as effective a contraceptive as twenty-eight-day-cycle Pills, its innovation with this product had nothing to do with contraception per se. Barr initially advertised Seasonale explicitly as a "lifestyle choice," designed to allow women who already took birth control anyway (and maybe also those who did not) to experience menstruation much less frequently. While it quickly backed away from the label "lifestyle choice," reacting to complaints that "lifestyle" enhancement was not a good reason to take prescription drugs, its advertising maintained the theme.[2]

What could be a more obvious extrapolation of the modern period than a product that could almost make menstruation disappear, without causing issues with delayed fertility or reversibility? By September of 2005, Barr could brag that more than a million prescriptions had been written and predicted that its product was well on its way to success.[3] As it turned out, however, Barr's product had a major hitch, and the many prescriptions written were not necessarily a good indicator of how well the product would do. Seasonale's potential Achilles' heel was that unpredictable breakthrough bleeding was a common occurrence that could continue for the entire first year of use. In fact, during the first year of use, women could expect to have as many bleeding days as on a regular Pill schedule, but distributed unevenly and occurring unpredictably, especially during the first six months.[4] If Seasonale was a "lifestyle" product, one supposed to facilitate modern menstrual management, this kind of breakthrough bleeding was certainly a major flaw.

In January of 2005, the FDA's Division of Drug Marketing, Advertising, and Communications informed Barr Pharmaceuticals that one of its television ads contained false and misleading advertising. The FDA found it problematic that Barr did not fully disclose the information about breakthrough bleeding contained in its product packages in its television ad, and that Barr's vague statement that women may "initially" experience breakthrough bleeding did not accurately represent the fact that many women would experience it for up to a year.[5] It is striking that the FDA saw the need to intervene for this reason. One might expect that the FDA would be most concerned with claims made about more clearly "medical" issues, such as contraceptive effectiveness or effects on endometriosis or PMS. But everyone recognized that the real promise of this Pill was in its potential to help women instantiate modern bodies, and that claiming it produced only four periods a year clearly implied that bleeding would be limited to these periods between active pills. Women in the initial trial, 7.7 percent of whom dropped out of the trial because of unacceptable breakthrough bleeding, agreed with the FDA that what Barr was really promising was exactly four predictable periods of bleeding a year, and it was lying in a substantial and meaningful way if its Pill did not deliver on this promise.[6]

In the end, young women who expect to be on birth control pills for a number of years in a row may find it worth it to deal with six months or a year's worth of unpredictable bleeding in order to have more years of infrequent and predictable periods. But whether extended-cycle oral contraceptives ultimately gain a large share of the market or remain a niche product, it seems clear that their value, and that of any other technological or informational innovations related to menstruation, will continue for some time to be measured against the requirements of the modern period.

ACKNOWLEDGMENTS

Over the past ten years, while researching and writing this book, I have accumulated many intellectual and personal debts. First, I want to acknowledge the seventy-five interviewees who gave so generously of their time and their memories. I wish I could thank them by name. Without them, this book would not have been possible, and indeed, their stories are what gives this history its meaning.

This research was made possible by the material support provided by a National Science Foundation Graduate Research Fellowship, a Whiting Fellowship in the Humanities, and a research grant from the Charles Warren Center for Studies in American History at Harvard University, as well as a Mellon Postdoctoral Fellowship at the Newhouse Center for the Humanities and Department of Women's Studies at Wellesley College.

I have had the opportunity to conduct research at several archives, and I appreciate the assistance of archivists Pat Gossel at the Smithsonian, Alexandra Briseno and Cheryl Chouiniere at the History Factory, Linda Clarke at BBDO, Harry Finley at the Museum of Menstruation, and fellow researcher Rich Lindstrom at the Gilbreth collection at Purdue University. Many thanks also go to Michael and Luci Cedrone, John and Ann Cedrone, and Rena Selya and Terry Cohen, who generously hosted me during my research travel. I had much assistance with my interviews as well. Research assistant Nancy Redd helped tremendously in recruiting interviewees. Grace Shen and Amy Wu spent many hours helping me with my Chinese. Kelly Freidenfelds transcribed many interviews, imbuing the transcripts with her subtle understanding of how to capture spoken words, and as a much-appreciated bonus, shared with me her insightful musings about the material.

At Harvard, the History of Medicine Working Group, and the Department of the History of Science more generally, provided friendly and knowledgeable arenas for discussion. I particularly appreciate the contributions of my dissertation readers, Katy Park and Mario Biagioli, my American history mentor Laurel Ulrich, as well as graduate student and faculty colleagues Conevery Bolton, Michele Murphy, Jeremy Greene, Kristen Haring, Rena Selya, Michael Gordin, Nick King, Debbie Weinstein, David Jones, David Barnes, and Charles Rosenberg. Allan Brandt's steadfast support and encouragement, along with his example of kind and attentive teaching and rigorous and accessible scholarship, have inspired my deepened commitment to academic teaching and research.

The Med Heads (History and Social Studies of Medicine and the Body Working Group) became my intellectual home at the University of California at Berkeley and provided a great deal more welcome feedback on this research. I especially thank Tom Laqueur, Leslie Reagan, Susan Zieger, Alastair Iles, Kate O'Neill, Warwick Anderson, Dorothy Porter, and Brian Dolan, and outside of Med Heads, Paula Fass and Mariane Ferme at Berkeley, and Adele Clarke at the University of California at San Francisco. Charles Rosenberg and Tom Laqueur read my completed dissertation and gave helpful direction for expanding and deepening my research and analysis. The two years I spent as a Mellon Postdoctoral Fellow at the Newhouse Center for the Humanities and the Department of Women's Studies at Wellesley College were crucial to the final drafting of the manuscript. I received wonderful moral support and intellectual company from the Newhouse Center fellows and my colleagues in Women's Studies, as well as members of Feminist Inquiry. I especially thank Newhouse Center director Tim Peltason and women's studies and history of medicine mentor Susan Reverby.

Conference presentations of bits and pieces of this book have provided opportunities for invaluable feedback and discussion as well. I particularly thank my colleagues at the American Association of the History of Medicine, the History of Science Society, and the Society for the History of Technology. As the book came to fruition, I appreciated the careful attention and assistance of Robert J. Brugger, Julia Ridley Smith, and the rest of the staff at the Johns Hopkins University Press, and the thorough and thoughtful comments provided by the anonymous reader.

I could not have completed this book without my family. My parents, John and Lucy Freidenfelds, have always set an example of curiosity, empathy, independent thinking, and perseverance, and I have looked to that example many times along the way. My brother, Jason Freidenfelds, has encouraged me with his enthusiasm for my research and his concrete editing suggestions. My sister, Kelly Freidenfelds, is my best girlfriend and my most trusted colleague, and the extended conversations we had about my research were vital in shaping my analysis. My husband, Felix Wu, is my partner in everything. His intellectual, spiritual, and material companionship has shaped this project from beginning to end. My children, Sebastian and Oliver, enrich my perspective on all of it.

My grandmother, Nina Freidenfelds, passed away before this book was completed, but I am glad I at least had the chance to try on my Ph.D. gown for her. She was the first person I ever formally interviewed and doing so made a big impression on me. I felt passionate about recording her life story, and I found that I believed strongly that the stories of "ordinary" women are worth hearing, remembering, and understanding. Now I find myself spending my life doing just that. I hope I can do it with her spirit of openness and generosity. This book is dedicated to her memory.

APPENDIX

Interview Method

This book relies heavily on the 75 in-depth interviews I conducted between 1999 and 2001 with Americans of a range of ages, primarily among African Americans in the rural South, white Americans in New England, and Chinese Americans in California, as well as a few people from outside these groups. Interviews are generally seen as a nontraditional source for historians, so an extended explanation of how I conducted and interpreted my interviews seems useful.

First, some further demographic description of the interviewees: I interviewed 17 white women and men from New England (15 women, 2 men); of these, the oldest interviewee was born before 1910, and the youngest after 1970. I interviewed 14 African Americans born in the rural South (10 women, 4 men); of these, the oldest interviewees were born before 1920, and the youngest after 1980. I interviewed 31 Chinese Americans who were part of families who had immigrated beginning in the 1960s and who resided in urban California (25 women, 6 men); the oldest interviewees were born before 1940, and the youngest were born after 1970. And I interviewed 13 others outside these groups, including those from Ashkenazi Jewish, Japanese, African American, Chinese American, Cape Verdean, and Protestant white backgrounds (12 women, 1 man); the oldest interviewees from this group were born before 1930, and the youngest were born after 1970.

Education is the best proxy I have for class and economic standing. Excluding Chinese Americans from immigrant families, of 15 interviewees born in 1940 or earlier, 14 had parents, some of whom were immigrants, with high school or less education (many with significantly less); one had a parent with a college education. Eight of them went to college, some later in their lives, and only one had less than a high school education. Of the 13 born between 1941 and 1960, seven had at least one parent who had gone to college, and 11 went to college themselves. Of the 14 interviewees born after 1960, 10 had at least one parent who had been to college, and 12 went to college themselves. My interviewees had more education than the average for their demographic groups, although their increasing levels of education over the century marked the same trend as can be seen for the entire U.S. population. For my oldest interviewees, their African American peers had on average 8.8 years of education, and their white peers on average 12.4 years. By 1960, African Americans had on average 12.7 years of education, and whites 12.8 years. The census stopped reporting these figures broken out by race for the youngest women and men I interviewed, but the average for the whole population had risen to 13.6 years.

The Chinese immigrants I interviewed came from well-off families, at least by Chinese standards (though most grew up with a standard of living in many ways significantly below their counterparts in the United States). They and their fathers, and sometimes mothers, were well educated, either in college or as military officers. All of the children of these immigrants I interviewed had gone or were going to college.

To obtain the interviews, I used what is often called the "snowball" method in anthropology and qualitative sociology: I started with someone I knew, in each of the ethnic/regional groups I interviewed, and then asked if she could refer me to someone else. With each interview I conducted, I requested this type of referral. This means that while I knew a few of the interviewees personally, and many of them knew one or two of the other interviewees, they were not members of any single community or network.

I interviewed women and men with a standard interview guide, but I allowed each interview to take its own shape, roaming over the questions in the order that was suggested by interviewees' stories and interests. Interviews lasted anywhere between about a half an hour for some of the men's interviews to close to three hours for some of the women. Men's interviews averaged about 40 minutes, and women's, about 90 minutes, including time spent gathering background information about family, education, school, work, and religion. I conducted as many interviews as possible in person, but all of the Southern African American interviews were done by phone, as well as several of the New England interviews. All interviews were recorded and either transcribed or notated in shorter form, with the explicit permission of the interviewees.

Interviews involved a lot of laughter. We laughed at different times for different reasons; sometimes it was sympathetic laughter, sometimes nervous. Because it was an important aspect of the interviews, it is noted in brackets in direct quotes from interviewees throughout the book. Women were often laughing at the ridiculousness of a situation they recalled, or occasionally the ridiculousness of spending an hour and a half talking about menstruation. They also sometimes laughed at my naive questions. Men often laughed with discomfort at discussing this topic, particularly with a women interviewer.

When I interviewed women, more often than not, laughing together reinforced the feeling that in relating and listening to stories, particularly stories that had not been related before, we were creating grounds for empathy between us as women. Situations that seemed embarrassing or difficult at the time could seem at least a little bit humorous in retrospect, especially when we acknowledged that all women probably have dealt with those situations and just never mentioned it to each other. Laughter became a bit more nervous when women were not sure I would be able to empathize. While telling and listening to stories about past embarrassments can establish empathy and support when it turns out that everyone suffered the same difficulties in silence, it can be a risky activity if it turns out that one's worst fear is true—that everyone else really did not have the same problem! As it turns out, while there were many variations, and a striking pattern of generational change, telling about and listening to women's and men's experiences of menstruation in the twentieth-century United States provides more obvious grounds for empathy than for distance. At the same time, changes in social and material culture, and bodily management, have been profound enough over the century that there is plenty of room for improving understanding of, and sympathy for, how women's and men's experiences of menstruation have changed over those years.

In interpreting my interviews, I needed to be sensitive to the important ways in which interviews differ from archival materials, even though I used both as historical sources. Interviews are retrospective. They contain interviewees' contemporary interpretations of past events, and they only contain what was memorable five, ten, and in some cases fifty or more years after the fact. These characteristics of interviews are both weaknesses and strengths. Documents recorded at the time of an event are, in some sense, more trustworthy (though always partial, perspectival, and usually extremely fragmentary), as they contain immediate impressions unblemished by forgetfulness and reinterpretation. On the other hand, women's retrospective interpretations of their experiences are as interesting to the historian as their understanding of those events at the time. In writing this history, I was interested in which aspects of women's experiences they found memorable years later. I also found it useful to have older women compare and contrast their experiences with what they saw their daughters and granddaughters doing. They were able to tell me which changes they found meaningful. It is their interpretations of the changes in their lives, as much as the changes themselves, that give this book its power.

Additionally, the kinds of experiences documented and interpreted in this book are especially partial and fragmentary in archival records, making interviews an indispensable source. What few, brief records women kept about their experiences of menstruation are almost all documents produced by highly educated white women. An experience such as menstruation, which for most women was private and mundane, is impossible for a historian to capture in any useful depth without interviews, especially when seeking the experiences and practices of those who were not well educated and were less likely to keep diaries or write letters prolifically.

In drawing upon the interviews, my approach does not assume the transparency of "experience" and its description but does insist on granting a reality to bodily experience that is not absolutely linked to particular representations. As Kathleen Canning proposes, I take to heart Joan Scott's argument that "experience" cannot be taken to be transparent, uncontestable evidence, outside of socially created identities and discourses, but I continue nevertheless to place importance on the project of sharing women's narratives of their experiences and work with an assumption of a more robust individual agency than Scott would perhaps grant. Scott advocates examining the discourses that make possible certain identities and experiences; Canning believes that it is also desirable to "untangle the relationships between discourses and experiences by exploring the ways in which subjects mediated or transformed discourses in specific historical settings."[1] In this book, I take Canning's approach, exploring the new discourses surrounding menstruation introduced in the early twentieth century, through which women and men's gender and class identities and experiences were shaped, but also examining how women and men chose among conflicting discourses, often with awareness of how engagement with certain discourses helped produce desired identities, and gradually transformed these discourses.

With interviewing comes questions of how much and what kind of data is sufficient. Historians are usually constrained by the content of the archives, that is, what has been written down and kept. On topics related to sex and reproduction, the archives are usually thin, since few people kept records about these private and often mundane aspects of their lives, and were even less likely to knowingly allow them to become public. Historians working in time periods for which interviews are impossible have come up with some tremendously creative ways to find useful data. In addition, they gauge the amount of speculation in their work to the available data and to the kinds of conclusions they are trying to draw. All analyses are to some degree speculative; historians almost never have "all" the data. At the same time, some kinds of analysis are inherently more speculative than others. For example, little data exists for certain time periods and locations, such as the Middle Ages in Europe. Historians take what documents they have and use them to full advantage, filling in the gaps with their best guesses, and acknowledging the degree of speculation and interpretation. In my case, the constraint is not the total potentially available interviews but the available time and resources for conducting them.

While 75 interviews with women and men of several different ethnic and regional backgrounds and a range of ages obviously cannot fully represent the experiences of all Americans, there were surprisingly strong threads of common experience running through these women's and men's stories. With the context provided by an analysis of popular culture based on published and archival sources, these common threads are robust enough to support some broad conclusions about Americans' experiences of their bodies in the twentieth century. Further research will certainly offer many refinements and modifications to my generalizations, especially for specific groups or historical moments. The generalizations are nevertheless worth making.

When I began this research, I expected to find that women and men from different ethnic, regional, and income groups would experience and talk about menstruation very differently from each other. I sought to show how the dominant paradigm in public sources, such as sex education books and advertisements, was challenged in multiple ways by individuals recounting their private experiences. Had this, indeed, been what I found, the number of interviews would have been irrelevant; the point would have been their range. What I found surprised me. My interviewees' narratives were closely related to these public sources, as I have documented, and were in most ways similar to each other despite socioeconomic differences. This makes the question of how many interviews it takes to support my conclusions more urgent.

On the face of it, 75 interviews may not seem like a lot for trying to draw broad conclusions about American experiences over a century. However, given that the natural expectation would have been that I would have found a great deal of variation among individuals, and even more discrepancies among ethnic and regional groups, the relative uniformity of these narratives ought to be regarded as a significant result. If the patterns I found were not actually compatible with widespread patterns in the society, what are the chances that the first 75 people I interviewed would have narrated these patterns with such consistency? The patterns are too pronounced, and too evenly spread through my interviews, to be dismissed as chance.

It is important to recognize that certain kinds of claims can be statistically reasonable to make with a limited set of interviews. With 75 interviews, this study is clearly not quantitative. Using the snowball method, I also clearly cannot claim that I used statistically valid sampling or that they are statistically representative of the population. However, I can make an important and statistically relevant claim about representation: statistically, it would be very unlikely to find the kind of robust and broad agreement among these interviewees about the major points I make based on the interviews if, in fact, they would not turn out to be representative of the next 100 or 1,000 interviews I could theoretically conduct. It is somewhat like rolling a die 75 times and finding that we rolled 6 most of the time. In that case, we could be quite sure that the die is loaded. If I were to make a claim like, "30% of Americans believed X about menstruation, 40% believe Y, and 30% were split among other beliefs," I would be clearly overreaching. Therefore, I have been careful not to make these kinds of statistical claims. I have used the interviews to do two things: first, to make large-scale generalizations about things for which I found particularly widespread agreement, and, second, to show how agreed-upon ideas and practices were put into play in the particular contexts of individuals' lives.

More interviews along the same lines would certainly increase the robustness of these conclusions and provide the opportunity to map them in more detail. More focused interviews with particular groups would likely provide the means to refine the contours of the differences among groups, as my chapter about tampons begins to do, and would suggest more subtle but important differences than my necessarily limited interviewing can illuminate. In light of my findings about the relationship between perceptions of being "modern" and being middle class, it would be particularly important to interview women and men who do not consider themselves to be either "modern" or "middle class" or both. I hope this book will inspire further inquiry.

NOTES

Introduction

1. See Marina Moskowitz, *Standard of Living: The Measure of the Middle Class in Modern America* (Baltimore: Johns Hopkins University Press, 2004) for an insightful and concrete description of how this occurred in several specific arenas. Also Ruth Schwartz Cowan, *More Work for Mother: The Ironies of Household Technology from the Open Hearth to the Microwave* (New York: Basic Books, 1983), 194–95.

2. Samuel Haber, *Efficiency and Uplift: Scientific Management in the Progressive Era 1890–1920* (Chicago: University of Chicago Press, 1964); Olivier Zunz, *Making America Corporate, 1870–1920* (Chicago: University of Chicago Press, 1990); Paula S. Fass, *Outside In: Minorities and the Transformation of American Education* (New York: Oxford University Press, 1989); Daniel Eli Burnstein, *Next to Godliness: Confronting Dirt and Despair in Progressive Era New York City* (Urbana: University of Illinois Press, 2006).

3. Fass, *Outside In;* William Leach, *Land of Desire: Merchants, Power, and the Rise of a New American Culture* (New York: Pantheon, 1993); Gary S. Cross, *An All-Consuming Century: Why Commercialism Won in Modern America* (New York: Columbia University Press, 2000); Roland Marchand, *Advertising the American Dream: Making Way for Modernity, 1920–1940* (Berkeley and Los Angeles: University of California Press, 1985); Lizabeth Cohen, *A Consumer's Republic: The Politics of Mass Consumption in Postwar America* (New York: Knopf, 2003); Andrea Tone, *The Business of Benevolence: Industrial Paternalism in Progressive America* (Ithaca, NY: Cornell University Press, 1997), 80–83; Adele Clarke, "Modernity, Postmodernity and Reproductive Processes Ca. 1890–1990 or, 'Mommy, Where Do Cyborgs Come from Anyway?'" in *The Cyborg Handbook*, ed. Chris Hables Gray (New York: Routledge, 1995).

4. Mary P. Ryan, *Mysteries of Sex: Tracing Women and Men through American History* (Chapel Hill: University of North Carolina Press, 2006), 207–14, 258.

5. In 1939, Gallup conducted its first poll in which it asked people about their perception of their social class. Respondents were asked, "To what social class in this country do you feel you belong—middle class, or upper, or lower?" 86% of respondents said that they were middle, upper middle, or lower middle class. Only 5% classified themselves as lower class. Gallup Organization, Question ID: USGALLUP.39-150.QA02. For more detail and

discussion of Gallup polls and the historiography of perception of social class, see the essay on sources.

6. Wendy Kline, *Building a Better Race: Gender, Sexuality, and Eugenics from the Turn of the Century to the Baby Boom* (Berkeley and Los Angeles: University of California Press, 2001); Joan Jacobs Brumberg, *Fasting Girls: The Emergence of Anorexia Nervosa as a Modern Disease* (Cambridge, MA: Harvard University Press, 1988).

7. Brumberg, *Fasting Girls*; Susan Bordo, *Unbearable Weight: Feminism, Western Culture, and the Body* (Berkeley and Los Angeles: University of California Press, 1993).

8. Michel Foucault, *Discipline and Punish: The Birth of the Prison* (New York: Vintage Books, 1995). An example of applying Foucault is Carole Spitzack, *Confessing Excess: Women and the Politics of Body Reduction* (Albany: SUNY Press, 1990).

9. Kline, *Building a Better Race*; Rosalyn Terborg-Penn, "African-American Women's Networks in the Anti-Lynching Crusade," in *Gender, Class, Race, and Reform in the Progressive Era*, ed. Noralee Frankel and Nancy S. Dye (Lexington: University Press of Kentucky, 1991); Mark Thomas Connelly, *The Response to Prostitution in the Progressive Era* (Chapel Hill: University of North Carolina Press, 1980), 143–46; Allan M. Brandt, *No Magic Bullet: A Social History of Venereal Disease in the United States since 1880*, expanded ed. (New York: Oxford University Press, 1987), 84–92; Thomas R. Pegram, *Battling Demon Rum: The Struggle for a Dry America, 1800–1933* (Chicago: Ivan R. Dee, 1998).

10. Joan Jacobs Brumberg, "'Something Happens to Girls': Menarche and the Emergence of the Modern American Hygienic Imperative," *Journal of the History of Sexuality* 4, no. 1 (1993): 99–127; Elizabeth Arveda Kissling, *Capitalizing on the Curse: The Business of Menstruation* (Boulder, CO: Lynne Rienner, 2006); Julie-Marie Strange, "The Assault on Ignorance: Teaching Menstrual Etiquette in England, c. 1920s to 1960s," *Social History of Medicine* 14, no. 2 (2001): 247–65; Barbara Brookes and Margaret Tennant, "Making Girls Modern: Pakeha Women and Menstruation in New Zealand, 1930–70," *Women's History Review* 7, no. 4 (1998): 565–81.

11. See the essay on sources for an extended discussion of the historiography of studies of the body and how this book is situated in terms of feminist, Foucauldian, and other postmodern interest in discipline, resistance, agency, and pleasure.

Chapter 1 • Before "Modern" Menstrual Management

1. By the same token, while the "modern" way of thinking about menstruation blossomed dramatically starting at the end of the nineteenth century, preliminary indicators of the development of a modern way to think about menstruation came as early as a century before they began to take root robustly in the popular culture. This chapter focuses on precedents for "traditional" ways of thinking about and managing menstruation; following chapters will trace ground laid during the nineteenth century which would support the blooming of "modern" approaches in the twentieth. This approach highlights interviewees' perception of a paradigm shift from an old to a new way of thinking in their own and their children's lives. At the same time, it should be recognized that traditional and modern approaches to menstruation overlapped for quite a long time in terms of individual women's and men's lives, if not in the larger historical picture. Furthermore, what interviewees

experienced as "traditional" did not encompass the entire range of pre-twentieth-century beliefs about and experiences of menstruation (for example, no one mentioned beliefs about the magic power of menstrual blood). This chapter addresses specifically the historical roots of early twentieth-century experience, not the entire pre-"modern" history of menstruation.

2. All interviewee names are pseudonyms.

3. See the essay on sources for a discussion of why these patterns are commonly perceived as "Victorian."

4. Monica H. Green, "Flowers, Poisons and Men: Menstruation in Medieval Western Europe," in *Menstruation: A Cultural History*, ed. Andrew Shail and Gillian Howie (New York: Palgrave Macmillan, 2005), 61–62; Monica Helen Green, "From 'Diseases of Women' To 'Secrets of Women': The Transformation of Gynecological Literature in the Later Middle Ages," *Journal of Medieval and Early Modern Studies* 30, no. 1 (2000): 7–11.

5. Katharine Park, *Secrets of Women: Gender, Generation, and the Origins of Human Dissection* (New York: Zone Books, 2006), 91–93, 97–103; Judith Walzer Leavitt, *Brought to Bed: Childbearing in America, 1750 to 1950* (New York: Oxford University Press, 1986).

6. Carroll Smith-Rosenberg has demonstrated the existence of social and affective networks that tied women together, often in more intimate and affectionate relationships than those with husbands, fathers, and sons, at least by the late eighteenth century and through the nineteenth century (Carroll Smith-Rosenberg, "The Female World of Love and Ritual: Relations between Women in Nineteenth-Century America," in *Disorderly Conduct: Visions of Gender in Victorian America*, ed. Carroll Smith-Rosenberg [Oxford: Oxford University Press, 1986]). These networks were often made tangible and organized by childbirth and child care.

7. Laura Gowing, *Common Bodies: Women, Touch, and Power in Seventeenth-Century England* (New Haven: Yale University Press, 2003).

8. William Buchan, *Domestic Medicine; or, the Family Physician: Being an Attempt to Render the Medical Art More Generally Useful, by Shewing People What Is in Their Own Power Both with Respect to the Prevention and Cure of Diseases. Chiefly Calculated to Recommend a Proper Attention to Regimen and Simple Medicines* (Edinburgh: Balfour Auld and Smellie, 1769). One scenario commonly cited through the nineteenth century was that a girl would see her first menses and be frightened, and the blood would be stopped up from the fright, causing health problems. This scenario had at least somewhat older roots; Duden cites an example from around 1730 where an analogous story is told about a man and his experience with a first hemorrhoidal bleeding, understood to be the male equivalent of menstruation. Barbara Duden, *The Woman beneath the Skin: A Doctor's Patients in Eighteenth-Century Germany* (Cambridge, MA: Harvard University Press, 1991), 224–25, n. 24.

9. Nancy F. Cott, "Passionlessness: An Interpretation of Victorian Sexual Ideology, 1790–1850," in *Women and Health in America: Historical Readings*, ed. Judith Walzer Leavitt (Madison: University of Wisconsin Press, 1984); Mary P. Ryan, *Mysteries of Sex: Tracing Women and Men through American History* (Chapel Hill: University of North Carolina Press, 2006), 88–100; Carroll Smith-Rosenberg, "Puberty to Menopause: The Cycle of Femininity in Nineteenth-Century America," in Smith-Rosenberg, *Disorderly Conduct*, 24–25; Marilyn R. Brown, "Images of Childhood," in *Encyclopedia of Children and Childhood*, ed.

Paula S. Fass (New York: Macmillan Reference USA, 2004), 454; Leavitt, *Brought to Bed*, 40–42.

10. Marie Jenkins Schwartz, *Birthing a Slave: Motherhood and Medicine in the Antebellum South* (Cambridge, MA: Harvard University Press, 2006), 102.

11. For a discussion of the vernacular sexual culture of the nineteenth century as a carryover from previous centuries, see Helen Lefkowitz Horowitz, *Rereading Sex: Battles over Sexual Knowledge and Suppression in Nineteenth-Century America* (New York: Vintage, 2003), chaps. 2 and 9.

12. Emil Novak, *The Woman Asks the Doctor* (Baltimore: Williams and Wilkins, 1935), 134.

13. Leslie J. Reagan, *When Abortion Was a Crime: Women, Medicine, and Law in the United States, 1867–1973* (Berkeley and Los Angeles: University of California Press, 1997), 24.

14. Vivian Nutton, "Humoralism," in *Companion Encyclopedia of the History of Medicine*, ed. W. F. Bynum and Roy Porter (New York: Routledge, 1993). As Shigehisa Kuriyama notes about bloodletting—a treatment central to humoral medical practice, and relevant to humoral understanding of menstruation—"Beliefs already dear to Galen still informed [diarist] Charles Waterton's devotion to bleeding some seventeen hundred years later [in 1871]" ("Interpreting the History of Bloodletting," *Journal of the History of Medicine and Allied Sciences* 50, no. 1 [1995]: 13).

15. Joan Cadden, *Meanings of Sex Difference in the Middle Ages: Medicine, Science, and Culture* (New York: Cambridge University Press, 1993), 19, 28–30.

16. Green, "Flowers, Poisons and Men," 60–61.

17. Cadden, *Meanings of Sex Difference in the Middle Ages*, 41–42.

18. Green, "From 'Diseases of Women' To 'Secrets of Women.'"

19. Cadden, *Meanings of Sex Difference in the Middle Ages*, 26–30; Michael Stolberg, "Menstruation and Sexual Difference in Early Modern Medicine," in Shail and Howie, *Menstruation: A Cultural History*, 91–94.

20. For a typical, and widely read, example of advice based on this understanding of menstruation, see "Of the Menstrual Discharge," in Buchan, *Domestic Medicine*.

21. Susan E. Klepp, "Colds, Worms, and Hysteria: Menstrual Regulation in Eighteenth-Century America," in *Regulating Menstruation: Beliefs, Practices, Interpretations*, ed. Etienne Van de Walle and Elisha P. Renne (Chicago: University of Chicago Press, 2001).

22. For example, see Bartholomew Parr, *The London Medical Dictionary; Including under Distinct Heads of Every Branch of Medicine, Viz. Anatomy, Physiology, and Pathology, the Practice of Physic and Surgery, Therapeutics, and Materia Medica; with Whatever Relates to Medicine in Natural Philosophy, Chemistry, and Natural History* (London: Johnson, 1809), 189.

23. Stolberg, "Menstruation and Sexual Difference in Early Modern Medicine," 98–99.

24. For example, Michael Ryan, *A Manual of Midwifery: Or, Compendium of Gynaecology and Paidonosology; Comprising a New Nomenclature of Obstetric Medicine, with a Concise Account of the Symptoms and Treatment of the Most Important Diseases of Women and Children, and the Management of the Various Forms of Parturition*, 1st American from the 3d London ed. (Burlington, VT: Smith and Harrington, 1835), 229.

25. Thomas Walter Laqueur, *Making Sex: Body and Gender from the Greeks to Freud* (Cambridge, MA: Harvard University Press, 1990), 207–27.

26. John C. Gunn, *Gunn's Domestic Medicine: A Facsimile of the First Edition* (1830; Knoxville: University of Tennessee Press, 1986), 297–98.

27. Edward B. Foote, *Plain Home Talk About the Human System: The Habits of Men and Women, the Causes and Prevention of Disease, Our Sexual Relations and Social Natures: Embracing Medical Common Sense Applied to Causes, Prevention, and Cure of Chronic Diseases* (New York: Murray Hill, 1874), 459–60.

28. Frederick Hollick, *The Diseases of Woman, Their Causes and Cure Familiarly Explained: With Practical Hints for Their Prevention, and for the Preservation of Female Health* (New York: Burgess, Stringer, 1847), 153.

29. Ibid., 155.

30. Cadden, *Meanings of Sex Difference in the Middle Ages*, 268.

31. Hermann Senator and Siegfried Kaminer, *Health and Disease in Relation to Marriage and the Married State* (New York: Rebman, 1904), 249.

32. Charles William Malchow, *The Sexual Life, Embracing the Natural Sexual Impulse, Normal Sexual Habits and Propagation, Together with Sexual Physiology and Hygiene* (St. Louis: C. V. Mosby Co., 1923), 161.

33. Hugh Northcote, *Christianity and Sex Problems* (Philadelphia: Davis, 1907), 125.

34. See, for instance, ongoing debate among authors throughout Senator and Kaminer, *Health and Disease in Relation to Marriage*.

35. Ibid.; Theodoor H. van de Velde and Stella Browne, *Ideal Marriage: Its Physiology and Technique* (New York: Random House, 1930).

36. Laura Klosterman Kidd, "Menstrual Technology in the United States, 1854–1921" (Ph.D. diss., Iowa State University, 1994), 122.

37. Susan Strasser, *Waste and Want: A Social History of Trash* (New York: Metropolitan Books, 1999), 170.

38. Ellen J. Buckland, *Now Science Helps Women Solve an Age-Old Problem* (Cellucotton Products Co., 1923), 11.

39. Strasser, *Waste and Want*, 199–201.

40. Use of pocketbooks became much more common among middle-class women in the 1920s, when women began to carry cosmetics to "freshen up" (a practice that created a controversial spectacle on many occasions, until it became accepted practice). Vincent Vinikas, *Hard Sell: American Hygiene in an Age of Advertisement* (Ames: Iowa State University Press, 1992), 56; Kathy Lee Peiss, *Hope in a Jar: The Making of America's Beauty Culture* (New York: Metropolitan Books, 1998), 186. This practice seems to have taken a generation to catch on more generally, especially among high school girls.

41. Most of the Chinese American women and men interviewed for this book were part of the large wave of immigration from Taiwan and Hong Kong to the United States in the 1960s and early 1970s, or were children of these immigrants. Most of those born before 1949 were born on the Mainland, and fled to Taiwan or Hong Kong because of the Japanese occupation and Communist takeover; most of those born later were children of those who fled. These immigrants were generally well educated, coming to the United States for graduate school or to seek employment after college, and they maintained a middle- or

upper-middle-class lifestyle in the United States. Most of them lived in several places in the United States before settling down in urban California.

Given this history of the interviewees, I use the term "China" to encompass the Chinese mainland, Taiwan, and Hong Kong, and I refer specifically to the *mainland* when I mean to exclude Taiwan and Hong Kong. Scholars of China have debated extensively whether, and in what time periods, Taiwan and Hong Kong should be considered a part of China. I found that the general menstrual health beliefs and practices and menstrual technology described by interviewees from various parts of China were quite similar, so I feel comfortable generalizing about a "China" that includes Taiwan and Hong Kong (with the recognition that Western medicine was introduced by different means, and as part of different ideology, on the mainland than it was in Taiwan and Hong Kong after 1949).

42. Sabine Wilms, "The Art and Science of Menstrual Balancing in Early Medieval China," in Shail and Howie, *Menstruation: A Cultural History;* Francesca Bray, *Technology and Gender: Fabrics of Power in Late Imperial China* (Berkeley and Los Angeles: University of California Press, 1997), 326–34; Charlotte Furth, "Blood, Body and Gender: Medical Images of the Female Condition in China: 1600–1850," *Chinese Science* 7 (1986): 43–66; Charlotte Furth and Shu-Yueh Ch'en, "Chinese Medicine and the Anthropology of Menstruation in Contemporary Taiwan," *Medical Anthropology Quarterly* 6, no. 1 (1992): 31–32.

43. Furth and Ch'en, "Chinese Medicine," 41–42.

Chapter 2 • The Modern Way to Talk about Menstruation

1. Hildagarde Esper, *Your Sex Questions Answered; What Every Modern Mother and Father Should Know* (Hollywood, CA: 1934), 22.

2. William Salmon, *Culpepper's Compleat and Experienc'd Midwife: In Two Parts. I. A Guide for Child-Bearing Women . . . Ii. Proper and Safe Remedies for the Curing All Those Distempers That Are Incident to the Female Sex . . . A Work Far More Perfect Than Any yet Extant, and Highly Necessary for All Surgeons, Midwives, Nurses, and Child-Bearing Women,* 7th ed. (London: printed, and sold by the booksellers, 1740), 143–52.

3. William Buchan, chap. 48, "Diseases of Women," in *Domestic Medicine; or, the Family Physician: Being an Attempt to Render the Medical Art More Generally Useful, by Shewing People What Is in Their Own Power Both with Respect to the Prevention and Cure of Diseases. Chiefly Calculated to Recommend a Proper Attention to Regimen and Simple Medicines* (Edinburgh: Balfour Auld and Smellie, 1769).

4. Sarah Stage, *Female Complaints: Lydia Pinkham and the Business of Women's Medicine* (New York: Norton, 1979), 45–63.

5. John C. Gunn, *Gunn's Domestic Medicine: A Facsimile of the First Edition* (1830; Knoxville: University of Tennessee Press, 1986), 295–302.

6. Stage, *Female Complaints; Private Textbook,* n.d., vol. 401, Microfilm M-79, Lydia E. Pinkham Medicine Company, Records, 1776–1968, Schlesinger Library, Radcliffe Institute, Harvard University, Cambridge, Massachusetts; M. Mattson, *Anatomical Chart with Explanation of the Uterus, Vagina, Ovaries, Intestinal Canal, &C : Theory of Reproduction or Origin of Human Life; Description and Use of Mattson's "Improved Irrigator," Termed the "Vagi-*

nal Irrigator," 7th ed. (New York: printed for the Mattson Syringe Company, 1871); Joseph Ralph, *A Domestic Guide to Medicine: By Which Individuals, Both Male and Female, Are Enabled to Treat Their Own Complaints on a Safe and Easy Principle* (New York: sold by the author, 1835), 124–27.

7. Janet Farrell Brodie, *Contraception and Abortion in Nineteenth-Century America* (Ithaca, NY: Cornell University Press, 1994), 6–7.

8. Ibid., 284–86.

9. Ibid., 106–35; Martha H. Verbrugge, *Able-Bodied Womanhood: Personal Health and Social Change in Nineteenth-Century Boston* (New York: Oxford University Press, 1988), 89, cites menstruation specifically as a lecture topic at the Boston Ladies' Physiological Institute in the 1880s.

10. Alexandra Lord, "'The Great Arcana of the Deity': Menstruation and Menstrual Disorders in Eighteenth-Century British Medical Thought," *Bulletin of the History of Medicine* 73, no. 1 (1999): 45–46; Bartholomew Parr, *The London Medical Dictionary; Including under Distinct Heads of Every Branch of Medicine, Viz. Anatomy, Physiology, and Pathology, the Practice of Physic and Surgery, Therapeutics, and Materia Medica; with Whatever Relates to Medicine in Natural Philosophy, Chemistry, and Natural History* (London: Johnson, 1809), 189.

11. Even before Bischoff's landmark 1843 experiment, some scientists had already begun to speculate that ovulation was linked to menstruation; for example, see Thomas Laycock, *A Treatise on the Nervous Diseases of Women; Comprising an Inquiry into the Nature, Causes, and Treatment of Spinal and Hysterical Disorders* (London: Longman, Orme, Brown, Green, and Longmans, 1840). For a description of Bischoff's experiment and an extended discussion of nineteenth-century views of ovulation and of menstruation as "heat," see Thomas Walter Laqueur, *Making Sex: Body and Gender from the Greeks to Freud* (Cambridge, MA: Harvard University Press, 1990), 207–27.

12. For a summary of these different viewpoints, see [No first name] Zinke, "Menstruation: Its Anatomy, Physiology, and Relation to Ovulation," *American Journal of Obstetrics* 24 (1891): 810; Mary Putnam Jacobi, *The Question of Rest for Women During Menstruation* (New York: G. P. Putnam's Sons, 1877); or Albert Henry Buck and Thomas Lathrop Stedman, *A Reference Handbook of the Medical Sciences Embracing the Entire Range of Scientific and Practical Medicine and Allied Science* (New York: W. Wood, 1900). For a twentieth-century example of the last viewpoint, which was the first to be abandoned, see Francis Marshall and William Jolly, "Contributions to the Physiology of Mammalian Reproduction. Part I.—the Oestrus Cycle in the Dog. Part Ii.—the Ovary as an Organ of Internal Secretion," *Philosophical Transactions of the Royal Society of London* 198B (1906): 99–121.

13. See Horrock's commentary appended to Walter Heape, "The Menstruation and Ovulation of Monkeys and the Human Female," *Obstetrical Society of London Transactions* 40 (1898): 161–74. See also Havelock Ellis, *The Evolution of Modesty: The Phenomena of Sexual Periodicity: Auto-Erotism* (Philadelphia: F. A. Davis, 1900). Ellis, an early sex researcher, acknowledged over and over again that most women insisted that they did not feel particularly sexually interested during menstruation, but he attributed this to social taboos and insisted that without the weight of the taboos women would experience estrus and therefore heightened sexual feeling during menstruation.

14. John Harvey Kellogg, *Ladies' Guide in Health and Disease: Girlhood, Maidenhood, Wifehood, Motherhood* (Battle Creek, MI: Good Health Pub. Co., 1891), 61–64, 176–78; John Harvey Kellogg, *Plain Facts for Old and Young* (Burlington, IA: I. F. Segner, 1890), 89.

15. Edward B. Foote, *Plain Home Talk About the Human System: The Habits of Men and Women, the Causes and Prevention of Disease, Our Sexual Relations and Social Natures: Embracing Medical Common Sense Applied to Causes, Prevention, and Cure of Chronic Diseases* (New York: Murray Hill, 1874), 456.

16. Jeffrey P. Moran, *Teaching Sex: The Shaping of Adolescence in the 20th Century* (Cambridge, MA: Harvard University Press, 2000); Regina Markell Morantz, "Making Women Modern: Middle-Class Women and Health Reform in 19th-Century America," in *Women and Health in America: Historical Readings*, ed. Judith Walzer Leavitt (Madison: University of Wisconsin Press, 1984), 353; Nancy Tomes, *The Gospel of Germs* (Cambridge, MA: Harvard University Press, 1998), 185–204; Paula S. Fass, *Outside In: Minorities and the Transformation of American Education* (New York: Oxford University Press, 1989), 13–35.

17. Evelyn Brooks Higginbotham, *Righteous Discontent: The Women's Movement in the Black Baptist Church, 1880–1920* (Cambridge, MA: Harvard University Press, 1993), 34–5, 190–229.

18. Ibid., 19–46; Jaqueline Jones, *Labor of Love, Labor of Sorrow: Black Women, Work, and the Family from Slavery to the Present* (New York: Basic Books, 1985), 96–99, 142–46; Mary P. Ryan, *Mysteries of Sex: Tracing Women and Men through American History* (Chapel Hill: University of North Carolina Press, 2006), 162–98, 207–14.

19. Patricia J. Campbell, *Sex Education Books for Young Adults, 1892–1979* (New York: R. R. Bowker Co., 1979).

20. Moran, *Teaching Sex*, 98–104.

21. Ibid., 105–7.

22. For example, Edith Belle Lowry, *Teaching Sex Hygiene in the Public Schools* (Chicago: Forbes and Co., 1914); Walter M. Gallichan, *A Textbook of Sex Education for Parents and Teachers* (Boston: Small, Maynard and Co., 1921).

23. Joan Jacobs Brumberg, *The Body Project: An Intimate History of American Girls* (New York: Random House, 1997), 46.

24. Campbell, *Sex Education Books*, 72.

25. Ellen J. Buckland, *Now Science Helps Women Solve an Age-Old Problem* (Cellucotton Products Co., 1923), 2, 7, 13 and 14.

26. Margaret A. Lowe, *Looking Good: College Women and Body Image, 1875–1930* (Baltimore: Johns Hopkins University Press, 2003), 103–33; Ryan, *Mysteries of Sex*, 214–22; Paula S. Fass, *The Damned and the Beautiful: American Youth in the 1920's* (New York: Oxford University Press, 1977).

27. Geo. H. Williamson, *Personal Hygiene for Women* (Kotex Co., 1928), 4 and 6.

28. *Preparing for Womanhood*, c. 1928, 3–4, folder 44, box 9-418, Kimberly-Clark Archives, The History Factory, Chantilly, Virginia (hereafter cited as K-C).

29. Mary Pauline Callender, *Marjorie May's Twelfth Birthday* (London: Kotex Ltd., 1929), 11.

30. Lloyd Arnold, *Health Facts on Menstruation* (Kotex Co., 1933), 2.

31. Laqueur, *Making Sex*, 211–12, and 94, n. 47 and 48; Carl Gottfried Hartman, *Time*

of Ovulation in Women: A Study on the Fertile Period in the Menstrual Cycle (Baltimore: Williams and Wilkins Co., 1936), 24–50, 57–60.

32. Arnold, *Health Facts on Menstruation*, 4–5.

33. Mary Pauline Callender, *Marjorie May Learns About Life* (Chicago: International Cellucotton Products Co., 1936), 8.

34. Personal Products Corp., *The Periodic Cycle*, c. 1938, foreword, available at www.mum.org/percyc1.htm (accessed Sept. 6, 2008); Personal Products Corp., *What a Trained Nurse Wrote to her Young Sister*, c. 1938, 1 and 3, available at www.mum.org/trainur2.htm (accessed Sept. 6, 2008).

35. Personal Products Corp., *The Periodic Cycle*, 4.

36. Marion Jones, *Kotex Educational Program*, Feb. 17, 1964, p. 3, folder 15, box 9-426, K-C.

37. Ad, "The Story of Menstruation," c. 1947, folder 12, box 9-423, K-C.

38. Thomas Heinrich and Bob Batchelor, *Kotex, Kleenex, Huggies: Kimberly-Clark and the Consumer Revolution in American Business* (Columbus: Ohio State University Press, 2004), 123.

39. "The Story of Menstruation" publicity materials reprinted from *The Catholic School Journal*, January 1957, folder 44, box 9-418, K-C.

40. Elaine Tyler May, *Homeward Bound: American Families in the Cold War Era* (New York: Basic Books, 1988).

41. Susan K. Freeman, *Sex Goes to School: Girls and Sex Education before the 1960s* (Champaign: University of Illinois Press, 2008).

42. Joanne J. Meyerowitz, *Not June Cleaver: Women and Gender in Postwar America, 1945–1960* (Philadelphia: Temple University Press, 1994).

43. Ryan, *Mysteries of Sex*, 228–38.

44. Jones, *Kotex Educational Program*, 9.

45. Heinrich and Batchelor, *Kotex, Kleenex, Huggies*, 121.

46. Woodward's authorship was kept secret from the public but mentioned in a Kimberly-Clark internal memo; see Jones, *Kotex Educational Program*, 2.

47. Heinrich and Batchelor, *Kotex, Kleenex, Huggies*, 88.

48. Jones, *Kotex Educational Program*, 2.

49. Warren Russell Johnson, *Human Sexual Behavior and Sex Education: Perspectives and Problems* (Philadelphia: Lea and Febiger, 1968), 67.

50. Joan Jacobs Brumberg, "'Something Happens to Girls': Menarche and the Emergence of the Modern American Hygienic Imperative," *Journal of the History of Sexuality* 4, no. 1 (1993): 99–127.

51. Fass, *Outside In*, 168.

52. Brumberg, "'Something Happens to Girls,'" 123, n. 53.

53. Ryan, *Mysteries of Sex*, 260–61, 279–81.

54. T. J. Mathews and Brady E. Hamilton, "Mean Age of Mother, 1970–2000," *National Vital Statistics Reports* 51, no. 1 (2002): 1–16.

55. The Life Cycle Center, *Getting Married* (Kimberly-Clark Corporation, 1975), 19. See Emily Martin, *The Woman in the Body: A Cultural Analysis of Reproduction* (Boston: Beacon Press, 1987), 52–53, advocating a similar alternative narrative.

56. Judy Blume, *Are You There, God? It's Me, Margaret* (New York: Dell, 1970), 96.
57. "Trauma-Rama," *Seventeen*, Dec. 1998, 12 and 16. It is notable that the embarrassment was about the bra, not the period.
58. Jen Maxwell, "Accidents Happen," *Seventeen*, Mar. 1998, 134.
59. "Trauma-Rama," *Seventeen*, June 1998, 24.

Chapter 3 • The Modern Way to Behave while Menstruating

1. Edward H. Clarke, *Sex in Education: Or, a Fair Chance for the Girls* (Boston: J. R. Osgood and Co., 1873); Cynthia Eagle Russett, *Sexual Science: The Victorian Construction of Womanhood* (Cambridge, MA: Harvard University Press, 1989), 112–25.
2. Carroll Smith-Rosenberg and Charles Rosenberg, "The Female Animal: Medical and Biological Views of Woman and Her Role in Nineteenth-Century America," *Journal of American History* 60 (1973): 332–56.
3. Jan Lewis and Kenneth A. Lockridge, "'Sally Has Been Sick': Pregnancy and Family Limitation among Virginian Gentry Women, 1780–1830," *Journal of Social History* 22 (1988–89): 5–19.
4. Linda Gordon, *The Moral Property of Women: A History of Birth Control Politics in America* (Chicago: University of Illinois Press, 2007), 86–104.
5. Julia Ward Howe, ed., *Sex and Education: A Reply to Dr. E. H. Clarke's "Sex in Education"* (1874; reprint, New York: Arno, 1972), 129; Azel Ames Jr., *Sex in Industry* (Boston: James R. Osgood, 1875).
6. Howe, *Sex and Education*, 94–95.
7. Ibid., 28, 54, and 135.
8. Mary Putnam Jacobi, *The Question of Rest for Women During Menstruation* (New York: G. P. Putnam's Sons, 1877).
9. John Harvey Kellogg, *Ladies' Guide in Health and Disease: Girlhood, Maidenhood, Wifehood, Motherhood* (Battle Creek, MI: Good Health Pub. Co., 1891), 182–83; Henry M. Lyman, *The Book of Health; a Reliable Family Physician Giving in Detail the Cause, Symptoms, Treatment and History of All Diseases of the Human Body and Plain Instructions for the Care of the Sick with Full Directions for Treating Emergency Cases* (Providence, RI: W. P. Mason, 1898), 888; Alice B. Stockham, *Tokology, a Book for Every Woman*, 29th ed. (Chicago: Sanitary Publishing Co., 1885), 253 and 257.
10. Stockham, *Tokology*, 253.
11. Margaret A. Lowe, *Looking Good: College Women and Body Image, 1875–1930* (Baltimore: Johns Hopkins University Press, 2003), 47–50.
12. Martha Verbrugge, "Gym Periods and Monthly Periods: Concepts of Menstruation in American Physical Education, 1900–1940," in *Body Talk: Rhetoric, Technology, Reproduction*, ed. Mary M. Lay et al. (Madison: University of Wisconsin Press, 2000); Russett, *Sexual Science*, 160–64.
13. Verbrugge, "Gym Periods," 79–83.
14. Clelia Duel Mosher, *Woman's Physical Freedom*, 3d rev. and enl. ed. (New York: The Womans Press, 1923), 22–31, 35.
15. Russett, *Sexual Science*, 161–62.

16. Paula S. Fass, *The Damned and the Beautiful: American Youth in the 1920's* (New York: Oxford University Press, 1977).

17. Stuart M. Blumin, *The Emergence of the Middle Class: Social Experience in the American City, 1760–1900* (Cambridge: Cambridge University Press, 1989), 290.

18. Mary P. Ryan, *Mysteries of Sex: Tracing Women and Men through American History* (Chapel Hill: University of North Carolina Press, 2006), 285; Jaqueline Jones, *Labor of Love, Labor of Sorrow: Black Women, Work, and the Family from Slavery to the Present* (New York: Basic Books, 1985), 91, 96–99, 169.

19. Andrea Tone, *The Business of Benevolence: Industrial Paternalism in Progressive America* (Ithaca, NY: Cornell University Press, 1997), 80–83.

20. Olivier Zunz, *Making America Corporate, 1870–1920* (Chicago: University of Chicago Press, 1990), 115–20.

21. Ruth E. Ewing, "A Study of Dysmenorrhea at the Home Office of the Metropolitan Life Insurance Company," *Journal of Industrial Hygiene* 13, no. 7 (1931): 244–51.

22. For example, a summary of C. L. Farrell, "Women in Industry," *Rhode Island Medical Journal* 26 (1943), cited in *Journal of Industrial Hygiene and Toxicology* 26 (1944): 303, among many others.

23. Margaret Castex Sturgis, "Observations on Dysmenorrhea Occurring in Women Employed in a Large Department Store," *Journal of Industrial Hygiene* 5, no. 2 (1923): 56.

24. Harry Walker Hepner, "Absenteeism of Women Office Workers," *Journal of Personnel Research* 3 (1924–25): 454–56.

25. Ellen J. Buckland, *Now Science Helps Women Solve an Age-Old Problem* (Cellucotton Products Co., 1923), 14.

26. Mary Pauline Callender, *Marjorie May's Twelfth Birthday* (Chicago: International Cellucotton Products Co., 1932), no page numbers.

27. Suellen M. Hoy, *Chasing Dirt: The American Pursuit of Cleanliness* (New York: Oxford University Press, 1995), 7, 22, 93, 123–28.

28. Ibid., 86, 88–89.

29. Eli F. Greer and Joseph H. Brown, *The New Tokology; Mother and Child Culture* (Chicago: Laird and Lee, 1903), 220; *As One Girl to Another* (International Cellucotton Products Co., 1940), 15.

30. Callender, *Marjorie May's Twelfth Birthday*.

31. Karl John Karnaky, "Vaginal Tampons for Menstrual Hygiene," *Western Journal of Surgery* 51 (1943): 152.

32. See cases in the medical literature as well: I. Lewis Sandler, "Dermatitis from a Sanitary Belt," *Medical Annals of the District of Columbia* 9, no. 1 (1940): 53; Frank E. Cormia, "Contact Dermatitis from Menstrual Pad," *JAMA* 107, no. 6 (1936): 429–30.

33. Emma Elizabeth Walker, *Beauty through Hygiene; Common Sense Ways to Health for Girls* (New York: A. S. Barnes Co., 1905), 180.

34. *Growing up and Liking It* (Milltown, NJ: Personal Products Corp., 1949), 7.

35. *You're a Young Lady Now* (Neenah, WI: Kimberly-Clark Corp., 1952), 5.

36. For example, Frederick A. Cleland, "A Method of Treatment of Severe Types of Dysmenorrhea with a Report of Results in 230 Cases," *American Journal of Obstetrics and Gynecology* 8 (1924): 337–45; A. J. Rongy, "The Use of X-Ray Therapy in Disturbed Men-

struation," in *Thirty-Sixth Annual Meeting of the American Association of Obstetricians, Gynecologists an Abdominal Surgeons* (Philadelphia: American Journal of Obstetrics and Gynecology, 1923); John Cooke Hirst, "The Comparative Value of Whole Ovarian Extract, Corpus Luteum Extract, and Ovarian Residue in Menstrual Disorder," *New York Medical Journal* (1921): 391–94.

37. By the late 1960s, nonsteroidal anti-inflammatory drugs more effective for menstrual cramps than aspirin were available by prescription. However, none of the women interviewed for this book remembered being prescribed Ibuprofen or Naproxen. These drugs became prominent in women's menstrual management only after they became available over the counter in the 1980s.

38. Beth L. Bailey, *Sex in the Heartland* (Cambridge, MA: Harvard University Press, 1999).

39. Dorothy Walter Baruch and Hyman Miller, *Sex in Marriage: New Understandings* (New York: Harper, 1962), 111; James Leslie McCary, *Human Sexuality: Physiological and Psychological Factors of Sexual Behavior* (New York: Van Nostrand Reinhold, 1967), 323; Allen Lein, *The Cycling Female, Her Menstrual Rhythm* (San Francisco: W. H. Freeman, 1979), 93.

40. *Ladies' Home Journal*, Feb. 1974, 20; *Ladies' Home Journal*, Oct. 1979, 63.

41. Katharina Dalton, *The Premenstrual Syndrome* (Springfield, IL: C. C. Thomas, 1964).

42. Robert T. Frank, "The Hormonal Causes of Pre-Menstrual Tension," *Archives of Neurological Psychiatry* 26 (1931): 1053–57; Dalton, *Premenstrual Syndrome*. Some examples of feminist sociology and anthropology are Sophie Laws, Valerie Hey, and Andrea Boroff Eagan, *Seeing Red: The Politics of Premenstrual Tension, Explorations in Feminism* (London: Hutchinson, 1985); Susan Markens, "The Problematic Of 'Experience': A Political and Cultural Critique of PMS," *Gender and Society* 10, no. 1 (1996): 53; Emily Martin, *The Woman in the Body: A Cultural Analysis of Reproduction* (Boston: Beacon Press, 1987), 113–38.

43. Michael Stolberg, "The Monthly Malady: A History of Premenstrual Suffering," *Medical History* 44 (2000): 312–15. Stolberg claims that the range of physical and emotional symptoms associated with PMS are evident as early as the late Renaissance, but he does not give convincing examples of emotional symptoms before the nineteenth century.

44. For example, Sandra S. Soria, "PMS: Battling the Monthly Blues," *Teen Magazine*, Dec. 1983, 24–26; Stephanie Young, "PMS Update: Practical Guidance from the Latest Research," *Glamour* 88, no. 4 (1990): 94–95; Joan K. Smith, "Eat to Beat PMS," *Redbook* 192, no. 5 (1999): 40; Michelle Meadows, "PMS Solutions: Calm the Storm with These Strategies," *Essence*, July 1998, 38.

Chapter 4 • The Modern Way to Manage Menstruation

1. *Cosmopolitan*, June 1937, 161.

2. T. J. Jackson Lears, "American Advertising and the Reconstruction of the Body, 1880–1930," in *Fitness in American Culture: Images of Health, Sport, and the Body, 1830–1940*, ed. Kathryn Grover (Amherst: University of Massachusetts Press, 1989), 56.

Notes to Pages 121–128   223

3. Nancy Tomes, *The Gospel of Germs* (Cambridge, MA: Harvard University Press, 1998), 62–66, 135–54.

4. *Ladies' Home Journal*, Apr. 1929, 53.

5. Olivier Zunz, *Making America Corporate, 1870–1920* (Chicago: University of Chicago Press, 1990); Susan Porter Benson, *Counter Cultures: Saleswomen, Managers, and Customers in American Department Stores, 1890–1940* (Urbana: University of Illinois Press, 1986).

6. Jane Farrell-Beck and Laura Klosterman Kidd, "The Roles of Health Professionals in the Development and Dissemination of Women's Sanitary Products, 1880–1940," *Journal of the History of Medicine and Allied Sciences* 51 (1996): 332–36.

7. Taylor Adams, "How Lasker Maneuvered for Kotex Ad Acceptance in 'Ladies' Home Journal,'" *Advertising Age* 2 (1974): 106–7.

8. Ronald H. Bailey, *Small Wonders: How Tambrands Began, Prospered and Grew* (Tambrands, Inc., 1986). Most of the ads explicitly cited below are from *Cosmopolitan*, but they are representative of those found in *Ladies' Home Journal, Good Housekeeping, Hygeia, Life*, etc.

9. Bailey, *Small Wonders*, 22–24.

10. See advertisements in *Printers Ink* promoting advertising space in magazines, for example, ad for *Hygiea*, June 17, 1949, 58; ad for *Mademoiselle*, Feb. 6, 1942, 71; ad for *Cosmopolitan*, June 6, 1947; ad for the Dell Modern Group (*Modern Screen, Modern Romances, Screen Stories*), Sept. 3, 1948.

11. Roland Marchand, *Advertising the American Dream: Making Way for Modernity, 1920–1940* (Berkeley and Los Angeles: University of California Press, 1985), 64.

12. *The Gallup Poll: Public Opinion* (Wilmington, DE: Scholarly Resources, 1972), 148.

13. *Ladies' Home Journal*, May 1929, 97.

14. *Cosmopolitan*, Mar. 1936, 127; July 1938, 121; July 1937, 154; Aug. 1948, 129.

15. *Cosmopolitan*, June 1941, 76; Aug. 1947, 125; Sept. 1936, 2.

16. *Cosmopolitan*, Mar. 1939, 73; Sept. 1938, 76; Sept. 1945, 116.

17. Lillian Gilbreth, "Report of Gilbreth, Inc.," Jan. 1, 1927, p. 63–64, folder 0704-2 NGRJ3, box 95, "N" series, Papers of Frank and Lillian Gilbreth, Special Collections, Purdue University Libraries, West Lafayette, Indiana.

18. *Cosmopolitan*, Sept. 1941, 111.

19. *Ladies' Home Journal*, Feb. 1929, 57; *Cosmopolitan*, Sept. 1945, 116; Sept. 1948, 96; Apr. 1947, 88.

20. Lears, "American Advertising," makes the argument that odor became a concern as people moved into cities (and therefore closer together), but this seems like an unsatisfactory explanation, since Europeans seemed to show so much less concern than Americans about the same process of urbanization.

21. Listerine ad, *Cosmopolitan*, Jan. 1948, 3; Packers Pine Tar Shampoo: "Don't let 'Scalp Odor' ruin romance," in *Cosmopolitan*, Oct. 1942, 112; Kleinert's Dress Shields, *Cosmopolitan*, Apr. 1941, 141; Mum deodorant, *Cosmopolitan*, Dec. 1938, 75. In addition to body deodorizers, home deodorizers were also advertised. They included not only cleaning products but also simple air fresheners. See ad for Air-wick, *Cosmopolitan*, Feb. 1950, 27.

22. *Ladies' Home Journal*, May 1901, 34–35; *Cosmopolitan*, Dec. 1938, 75.

23. *Cosmopolitan*, Dec. 1943, 174; Apr. 1950, 103; Jan. 1948, 3.

24. *Cosmopolitan*, Sept. 1936, 177; Mar. 1937, 179; May 1939, 142.

25. *Cosmopolitan*, Nov. 1948, 92; Mar. 1937, 179.

26. Warren Susman, *Culture as History: The Transformation of American Society in the Twentieth Century* (New York: Pantheon, 1984), chap. 14; Elizabeth Haiken, *Venus Envy: A History of Cosmetic Surgery* (Baltimore: Johns Hopkins University Press, 1997), 100.

27. *Cosmopolitan*, Feb. 1950, 119.

28. *Cosmopolitan*, Feb. 1940, 105; *Hygeia*, Sept. 1943, 685; *Cosmopolitan*, Nov. 1939, 77.

29. Kathy Lee Peiss, *Cheap Amusements: Working Women and Leisure in New York City, 1880 to 1920* (Philadelphia: Temple University Press, 1985), 88–114; Fass, *Damned and the Beautiful*, 191–208.

30. *Ladies' Home Journal*, Nov. 1939, 44; Sharon Hartman Strom, *Beyond the Typewriter: Gender, Class, and the Origins of Modern American Office Work, 1900–1930* (Urbana: University of Illinois Press, 1992).

31. Fass, *Damned and the Beautiful*, 137, 234.

32. *Cosmopolitan*, June 1938, 124; Sept. 1939, 94; June 1949, 82.

33. Ibid., Oct. 1940, 129.

34. Ibid., July 1938, 105: "Housewives, office workers, college girls, sports lovers—all are adopting" Tampax.

35. *Cosmopolitan*, June 1942, 9, ad for Kotex showing women in various uniforms: nurse, factory, and army. In April 1944 came the last of explicit reference to women doing blue-collar work in *Cosmopolitan*, though there were still occasional women in military uniforms in Kotex ads, and references to the war continued, though less prominently. There is a lone reference to war work in a 1945 Midol ad (*Cosmopolitan*, Mar. 1945, 118).

36. One ad, in 1944, does mention gardeners and taxi drivers; taxi drivers were presumably part of the war effort, and gardeners may have been a reference to women who created "victory gardens" to help feed their families. *Cosmopolitan*, Apr. 1944, 106. For a direct reference to victory gardens, with illustration, see Kotex ad, *Cosmopolitan*, May 1944, 9.

37. *Cosmopolitan*, Apr. 1944, 131; June 1944, 131; Aug. 1944, 79; Sept. 1944, 159; June 1946, 132; Aug. 1946, 150; Oct. 1946, 80.

38. For these examples and more of mothers in ads, see *Ladies Home Journal*, Feb. 1929, 151, 168, 173 and 179; Mar. 1929, 58–59.

39. Lears, in "American Advertising," says that ads promoted a certain kind of cultural imperialism by insisting that all people should embrace the "civilized" use of soap, etc.; at the same time, he also says that images of "hairless bodies and sterile households" (60) provided a means of setting white elites apart from immigrants, blacks, and others in a time of cultural turmoil. I interpret these facts differently: manufacturers and advertisers were willing to sell the products necessary to realize a middle-class body to anyone who could pay, and they told ambitious newcomers as well as established elites that they could perform middle-class identity (involving particular activities, personality, self-presentation, and maybe also race/ethnicity) if they only had the right products. (And if they didn't use the right products, they risked being pegged as something other than middle class.) While the purchasers of products were trying to set themselves apart from the masses, or at least

make sure they were members of the "right" masses, by realizing middle-class bodies and identities, the producers of the products had a much more "imperialist" goal, trying to sell their products to as many people as possible.

40. Calculated according to the Consumer Price Index for 2000.

41. Farrell-Beck and Kidd, "The Roles of Health Professionals," 332–36.

42. Susan Strasser, *Waste and Want: A Social History of Trash* (New York: Metropolitan Books, 1999).

43. Ida Smithson was characteristic of many African American mothers, in her hard work, self-sacrifice, and educational ambitions for her children. See Jaqueline Jones, *Labor of Love, Labor of Sorrow: Black Women, Work, and the Family from Slavery to the Present* (New York: Basic Books, 1985), 96–99.

44. Joan Jacobs Brumberg, *The Body Project: An Intimate History of American Girls* (New York: Random House, 1997), 44.

45. Ruth Schwartz Cowan, *More Work for Mother: The Ironies of Household Technology from the Open Hearth to the Microwave* (New York: Basic Books, 1983), 194–95.

46. R. W. Ebert, Brand Manager, "History of Kotex Feminine Napkins, 1914–1969," folder 18, box 9-12, Kimberly-Clark Archives, The History Factory, Chantilly, Virginia (hereafter cited as K-C).

47. Sharra Louise Vostral, "Conspicuous Menstruation: The History of Menstruation and Menstrual Hygiene Products in America, 1870–1960" (Ph.D. diss., Washington University, 2000), 135–37.

48. For example, Shelley M. Park, "From Sanitation to Liberation?: The Modern and Postmodern Marketing of Menstrual Products," *Journal of Popular Culture* 30, no. 2 (1996): 149–69; Rebecca Ginsberg, "'Don't Tell, Dear': The Material Culture of Tampons and Napkins," *Journal of Material Culture* 1, no. 3 (1996): 365–75; Janice Delaney, Mary Jane Lupton, and Emily Toth, *The Curse: A Cultural History of Menstruation* (Chicago: University of Illinois Press, 1988), 132.

49. Ebert, "History of Kotex," 1922–23, p. 3; 1924, p. 4; 1928, p. 1.

50. Ibid., 1934, p. 1.

51. *As One Girl to Another* (International Cellucotton Products Co., 1940), 8.

52. Ebert, "History of Kotex," 1937–38, p. 4; 1951, p. 6.

53. Ibid., 1954, p. 5.

54. "Kotex Packaging," Dec. 16, 1958, folder 8, box 9-357, K-C.

55. Ibid., 10.

56. Strasser, *Waste and Want*, 167.

57. Ebert, "History of Kotex," 1937–38, p. 6.

58. Susan Strasser, *Satisfaction Guaranteed: The Making of the American Mass Market* (New York: Pantheon, 1989), chap. 2.

59. "Report of Gilbreth, Inc.," p. 60, 62.

60. Ebert, "History of Kotex," 1936, p. 1.

61. Strasser, *Waste and Want*, 164.

62. Ebert, "History of Kotex," 1937–38, p. 3.

63. Theodore Caplow, Louis Hicks, and Ben J. Wattenberg, *The First Measured Century: An Illustrated Guide to Trends in America, 1900–2000* (Washington, DC: AEI Press, 2001), 99.

64. Marina Moskowitz, *Standard of Living: The Measure of the Middle Class in Modern America* (Baltimore: Johns Hopkins University Press, 2004), 64–104.

65. "Report of Gilbreth, Inc.," 63–64.

66. This pattern of change and distancing between generations is also evident in Gertrude Jacinta Fraser's account of rural African Americans' explanations of how and why their health beliefs and practices changed over the twentieth century ("Modern Bodies, Modern Minds: Midwifery and Reproductive Change in an African American Community," in *Conceiving the New World Order: The Global Politics of Reproduction,* ed. Faye D. Ginsburg and Rayna Rapp [Berkeley and Los Angeles: University of California Press, 1995]).

67. Evelyn Brooks Higginbotham, *Righteous Discontent: The Women's Movement in the Black Baptist Church, 1880–1920* (Cambridge, MA: Harvard University Press, 1993), 34–5, 190–229, 188–94.

68. Ironically, o.b. carved out a niche for itself by selling tampons without applicators as a fantastic innovation in tampon technology. In fact, nonapplicator tampons came out at the very beginning but lost most of the market to Tampax because for most of the century, women preferred tampons with applicators.

69. For a fascinating history of Pill dial technology, see Patricia Peck Gossel, "Packaging the Pill," in *Manifesting Medicine: Bodies and Machines,* ed. Robert Bud, Bernard S. Finn, and Helmuth Trischler (Amsterdam: Harwood Academic, 1999).

Chapter 5 • Tampons

1. J. Milton Singleton and Herbert F. Vanorden, "Vaginal Tampons in Menstrual Hygiene," *Western Journal of Surgery* 51 (1943): 146–49.

2. Lloyd Arnold and Marie Hagele, "Vaginal Tamponage for Catamenial Sanitary Protection," *Journal of the American Medical Association* 110 (1938): 791.

3. Harry Sackren, "Vaginal Tampons for Menstrual Absorption," *Clinical Medicine and Surgery* 46 (1939): 327–29.

4. Maurice O. Magid and Jacob Geiger, "The Intravaginal Tampon in Menstrual Hygiene," *Medical Record* 155 (1942): 316–20; Karl John Karnaky, "Vaginal Tampons for Menstrual Hygiene," *Western Journal of Surgery* 51 (1943): 150–52.

5. Singleton and Vanorden, "Vaginal Tampons in Menstrual Hygiene."

6. Arthur Hale Curtis, *A Textbook of Gynecology* (Philadelphia: W. B. Saunders Company, 1942), 122; Arthur Hale Curtis and John William Huffman, *A Textbook of Gynecology* (Philadelphia: W. B. Saunders Co., 1950), 129.

7. Arnold and Hagele, "Vaginal Tamponage for Catamenial Sanitary Protection," 790–92.

8. Sackren, "Vaginal Tampons for Menstrual Absorption;" Magid and Geiger, "The Intravaginal Tampon in Menstrual Hygiene."

9. Singleton and Vanorden, "Vaginal Tampons in Menstrual Hygiene," 149.

10. Ibid., 146. Apparently, tampons didn't inspire these temptations in married women; they presumably had their sexual desires met by their husbands.

11. Robert Latou Dickinson, "Tampons as Menstrual Guards," *JAMA* 128 (1945): 490.

12. Ibid., 492.

13. Compare "Sanitary Napkins," *Consumer Reports* 2 (1937): 24, with "Sanitary Pads and Tampons," *Consumer Reports* 10 (1945): 240, and Robert Latou Dickinson, "Tampons as Menstrual Guards," *Consumer Reports* 10 (1945): 246–47.

14. Museum of Menstruation material culture collection. Available at www.mum.org (accessed Sept. 6, 2008).

15. *As One Girl to Another*, folder 6, box 9-416, Kimberly-Clark Archives, The History Factory, Chantilly, Virginia (hereafter cited as K-C); "You can wear a Meds tampon when the vaginal opening is large enough to permit you to insert it easily. This is generally true when you are fully grown—toward the end of your teens. Consult your doctor," from Modess/Meds, *Growing up and Liking It* (Milltown, NJ: Personal Products Corp., 1949), folder 15, box 9-416, K-C.

16. Minutes of meeting held Thursday, July 31, 1952: Kotams, p. 6, folder 3, box 9-357, K-C.

17. Beth L. Bailey, *Sex in the Heartland* (Cambridge, MA: Harvard University Press, 1999).

18. Melanie Mannarino, "Sex + Body," *Seventeen*, Nov. 1998.

19. Joan Jacobs Brumberg, *The Body Project: An Intimate History of American Girls* (New York: Random House, 1997).

20. *Ladies' Home Journal*, Jan. 1984, 96.

21. Virginia L. Olesen, "Analyzing Emergent Issues in Women's Health: The Case of the Toxic-Shock Syndrome," in *Culture, Society, and Menstruation*, ed. Virginia L. Olesen and Nancy Fugate Woods (Washington, DC: Hemisphere Pub. Corp., 1986), 53–56.

22. According to a Kimberly-Clark memo, "The tampon share of the total feminine care market shifted from a historical 47 percent to 36 percent for the January/February, 1981 bimonthly period" ("Kotex Security Tampons Five Year Plan," May 5, 1981, p. 1–2, folder 103, box 9-341, K-C). While the author of this memo projected that sales would bounce back by late 1981 as publicity about TSS subsided (p. 2), in fact, the trend continued. Between 1978 and 1984, as TSS persisted as a topic in the news, the percentage of women using tampons dropped from 67% to 56%, the percentage using tampons exclusively dropped from 31% to 11%, and tampons' share of the feminine care market dropped from 47% to 34% ("Tampon Update," Sept. 25, 1984, folder 144, box 9-346, K-C). By 2001, a poll conducted by another company concluded that 37% of American women over the age of eighteen used tampons, skewed heavily toward young women ("The U.S. Market for Women's OTC Reproductive Healthcare Products," Report LA496 (1997) by Packaged Facts, 625 Avenue of the Americas, New York, NY). Concern about TSS coincided with the development of significantly thinner, more absorbent and more reliable pads, which seems to have persuaded many women, especially older women, to forgo tampons completely. It also seems likely that the two polling organizations were measuring use of tampons differently; Kimberly-Clark seems to have measured whether women ever used tampons at all, and the Packaged Facts description does not specify.

Conclusion

1. For example, Emily Martin, *The Woman in the Body: A Cultural Analysis of Reproduction* (Boston: Beacon Press, 1987), 54–68; Rose Weitz, *The Politics of Women's Bodies: Sexuality, Appearance, and Behavior* (New York: Oxford University Press, 1998), Sharlene Janice Hesse-Biber, *Am I Thin Enough Yet?: The Cult of Thinness and the Commercialization of Identity* (New York: Oxford University Press, 1996).

2. Elizabeth Arveda Kissling, *Capitalizing on the Curse: The Business of Menstruation* (Boulder, CO: Lynne Rienner, 2006), 67.

3. "Barr Pharmaceuticals, Inc.; Approvable Letter Issued for Seasonique Extended-Cycle Oral Contraceptives," *Patient Care Law Weekly*, Sept. 18, 2005, 26.

4. In a trial conducted to obtain FDA approval, researchers followed 682 women who took either the extended cycle Pill or the typical twenty-eight-day Pill. They found that it took more than three-quarters of a year for the median number of breakthrough bleeding days to be about the same for extended-cycle users as for regular Pill users. During the first three-month cycle, half of the women on the extended-cycle Pill experienced twelve or more days of breakthrough bleeding or spotting. This figure gradually reduced to four days during the fourth three-month cycle, similar to the results with the conventional twenty-eight-day Pill. Ultimately, 7.7 percent of the women taking the extended cycle oral contraceptive dropped out of the study because of unacceptable bleeding, compared to 1.8 percent of women taking the conventional Pill. F. D. Anderson, Howard Hait, and the Seasonale-301 Study Group, "A Multicenter, Randomized Study of an Extended Cycle Oral Contraceptive," *Contraception* 68 (2003): 89–96.

5. Inc. Washington Business Information, "Barr Told to Halt TV Ad Promoting Seasonale," *Pharmaceutical Corporate Compliance Report*, Jan. 17, 2005.

6. It is difficult to evaluate just how successful or unsuccessful extended-cycle oral contraceptives will eventually be. As of the beginning of 2007, they appeared to have enough potential that one of Barr Pharmaceutical's competitors had brought out a generic version of Seasonale, and Barr had found it worth responding by developing a second-generation extended-cycle Pill, Seasonique, as well as their own generic version of Seasonale. In May of 2007 Wyeth received FDA approval for Lybrel, a birth control pill that eliminated withdrawal bleeding completely. At the same time, investment analysts were disappointed with the performance of Barr's extended-cycle products, enough to name this as a significant reason to downgrade the value of Barr's stocks. Robert Uhl and Alan Meyers, "Barr Pharmaceuticals Inc." (FBR Research, 2006).

Appendix

1. Kathleen Canning, "Feminist History after the Linguistic Turn: Historicizing Discourse and Experience," *Signs* 19, no. 2 (1994): 373; Joan Scott, "Experience," in *Feminists Theorize the Political*, ed. Judith P. Butler and Joan Wallach Scott (New York: Routledge, 1992).

ESSAY ON SOURCES

The first major burst of academic writing on the social and cultural history of menstruation began in the 1970s. Much of the scholarly writing of the 1970s and 1980s was focused on the nineteenth century: Carroll Smith-Rosenberg, "Puberty to Menopause: The Cycle of Femininity in Nineteenth-Century America," *Feminist Studies* 1, no. 3/4 (1973): 58–72; Elaine Showalter and English Showalter, "Victorian Women and Menstruation," in *Suffer and Be Still: Women in the Victorian Age*, ed. Martha Vicinus (Bloomington: Indiana University Press, 1972); Vern Bullough and Martha Voght, "Women, Menstruation, and Nineteenth-Century Medicine," in *Women and Health in America: Historical Readings*, ed. Judith Walzer Leavitt (Madison: University of Wisconsin Press, 1984); Patricia Vertinsky, "Exercise, Physical Capability, and the Eternally Wounded Woman in Late Nineteenth Century North America," *Journal of Sport History* 14, no. 1 (1987): 7–27.

Because these path-breaking authors examining the relationship between social structure and menstrual beliefs and practices focused on the nineteenth century, many following them have assumed that the "traditional" menstrual beliefs described in chapter 1 of this book were the result of "Victorian" modesty and sexual repression. More recent scholarship on menstruation in earlier time periods has demonstrated the deeper historical roots of these patterns, without merely pointing to the menstrual regulations in Leviticus or beliefs about dangers of menstrual blood and menstruating women as compiled by Pliny; see essays in Andrew Shail and Gillian Howie, *Menstruation: A Cultural History* (New York: Palgrave Macmillan, 2005), especially Monica H. Green, "Flowers, Poisons and Men: Menstruation in Medieval Western Europe;" Joan Cadden, *Meanings of Sex Difference in the Middle Ages: Medicine, Science, and Culture* (New York: Cambridge University Press, 1993); and the historical essays in Etienne Van de Walle and Elisha P. Renne, eds., *Regulating Menstruation: Beliefs, Practices, Interpretations* (Chicago: University of Chicago Press, 2001). For a discussion of the vernacular sexual culture of the nineteenth century as a carryover from previous centuries, see Helen Lefkowitz Horowitz, *Rereading Sex: Battles over Sexual Knowledge and Suppression in Nineteenth-Century America* (New York: Vintage, 2003), chapters 2 and 9.

The impressively long-lived nature of pre-"modern" menstrual beliefs in Western culture does not, of course, mean that we should interpret them as "natural" or universal. For cross-cultural examples of the variety of beliefs and practices surrounding menstruation and a sophisticated analysis of the concept of menstrual "taboo," see Thomas Buckley and Alma Gottlieb, eds., *Blood Magic: The Anthropology of Menstruation* (Berkeley and Los Angeles: University of California Press, 1988).

In-depth examinations of menstruation for a popular audience, imbued with a consciousness-raising spirit, emerged out of the feminist movement of the 1970s as well—for example, Janice Delaney, Mary Jane Lupton, and Emily Toth, *The Curse: A Cultural History of Menstruation* (New York: Dutton, 1976); Paula Weideger, *Menstruation and Menopause: The Physiology and Psychology, the Myth and the Reality* (New York: Knopf, 1976). While these books were popular compared to scholarly writings, they were not mentioned by my interviewees who were old enough to read these books in the 1970s or 1980s, even when prompted to think about whether they had read anything "feminist" about menstruation. They are perhaps reflected in the experiences of some of the youngest interviewees, who had read feminist anthropological accounts of menstruation by scholars who may well have been inspired by these earlier feminist works.

Another important strand of research on menstruation has specifically examined menstrual technology and its promotion. Among the first is Fred E. H. Schroeder, "Feminine Hygiene, Fashion, and the Emancipation of American Women," *American Studies* 17, no. 2 (1976): 101–10, which argues that the technological innovation of disposable menstrual pads was a necessary precursor to the shortening of skirts and the growth of women's participation in the workplace in the early twentieth century. Schroeder seems to see technological innovation driving social change; he does not discuss homemade disposable pads, an innovation that did not require any newly available technology or frankness in advertising. Anne M. Spurgeon, "Marketing the Unmentionable: Wallace Meyer and the Introduction of Kotex," *Maryland Historian* 19, no. 1 (1988): 17–30, describes the first Kotex ad campaign, and its creator's emphasis on "modern" womanhood. Laura Klosterman Kidd, "Menstrual Technology in the United States, 1854–1921" (Ph.D. diss., Iowa State University, 1994) examines the pre-Kotex material culture history of menstrual technology.

Jane Farrell-Beck and Laura Klosterman Kidd, "The Roles of Health Professionals in the Development and Dissemination of Women's Sanitary Products, 1880–1940," *Journal of the History of Medicine and Allied Sciences* 51 (1996): 325–52, points out that advertisers were not the only promoters of commercial menstrual technology. Susan Strasser, *Waste and Want: A Social History of Trash* (New York: Metropolitan Books, 1999) spends a chapter describing Kotex as one of the first major products of the new culture of disposability, which complemented the rising culture of consumption. Thomas Heinrich and Bob Batchelor, *Kotex, Kleenex, Huggies: Kimberly-Clark and the Consumer Revolution in American Business* (Columbus: Ohio State University Press, 2004) gives the business side of the technological history.

Much excellent work has been done on the medical and cultural history of menstruation in China before the twentieth century: Charlotte Furth, "Blood, Body and Gender: Medical Images of the Female Condition in China: 1600–1850," *Chinese Science* 7 (1986): 43–66; Francesca Bray, *Technology and Gender: Fabrics of Power in Late Imperial China* (Berkeley and Los Angeles: University of California Press, 1997); Sabine Wilms, "The Art and Science of Menstrual Balancing in Early Medieval China," in *Menstruation: A Cultural History*, ed. Andrew Shail and Gillian Howie (New York: Palgrave Macmillan, 2005). Resources are sparser for the twentieth century, but see Charlotte Furth and Shu-Yueh Ch'en, "Chinese Medicine and the Anthropology of Menstruation in Contemporary Taiwan," *Medical Anthropology Quarterly* 6, no. 1 (1992): 27–48.

As described in the introduction, much of more recent historical scholarship on menstruation in the twentieth-century United States has analyzed this history in feminist and Foucauldian terms, attentive to social control, discipline, and inculcation of bodily norms (for example, Joan Jacobs Brumberg, "'Something Happens to Girls': Menarche and the Emergence of the Modern American Hygienic Imperative," *Journal of the History of Sexuality* 4, no. 1 (1993): 99–127; Elizabeth Arveda Kissling, *Capitalizing on the Curse: The Business of Menstruation* (Boulder, CO: Lynne Rienner, 2006); Martha Verbrugge, "Gym Periods and Monthly Periods: Concepts of Menstruation in American Physical Education, 1900–1940," in *Body Talk: Rhetoric, Technology, Reproduction*, ed. Mary M. Lay et al. (Madison: University of Wisconsin Press, 2000) focuses on how physical education teachers asserted new authority over girls' bodies in claiming expertise about menstruation and appropriate physical practices).

Anthropologist Emily Martin takes what I would describe as a feminist Marxist perspective, observing how the nineteenth and twentieth century "scientific" explanation of menstruation I describe in chapter 2 relies on industrial metaphors of production to explain reproductive processes, and depicts menstruation as a failure to produce a baby (*The Woman in the Body: A Cultural Analysis of Reproduction* [Boston: Beacon, 1987]). In interviews conducted in Philadelphia in the early 1980s, she finds a class difference among young women not evident in my interviews, which she interprets as middle-class acquiescence to, and working-class resistance to, this metaphor of failed production. Taken together, this scholarship represents an important thread in the "history of the body," a historical subfield newly self-consciously defined in the past thirty years but in many ways continuing work previously done in social history, women's history, medical history, etc. (For an overview of the first generation of scholarship in this newly defined area, see Barbara Duden, "A Repertory of Body History," in *Fragments for a History of the Human Body*, ed. Michel Feher, Ramona Naddaff, and Nadia Tazi [New York: Zone, 1989]).

The approach that has been primarily taken in writing the history of menstruation does not represent the only perspective on the history of the body, however. Feminist interest in "agency" and postmodern attention to "resistance," play, and pleasure have encouraged some historians to look outside a social control model, and some have looked at women's narratives of their experiences through this lens. For example, Kathy Lee Peiss has shown that white and African American women used cosmetics for pleasure, community-building, and entrepreneurial opportunities (*Hope in a Jar: The Making of America's Beauty Culture* [New York: Metropolitan Books, 1998]). Elizabeth Haiken has shown that even cosmetic surgery had positive elements of play, pleasure, and self-fashioning (Elizabeth Haiken, *Venus Envy: A History of Cosmetic Surgery* [Baltimore: Johns Hopkins University Press, 1997]). In the case of cosmetics and cosmetic surgery, these scholars acknowledge that these new beauty practices did have their oppressive aspects or modes of use; they ask not that the practices as a whole be lauded for improving women's lives but that the positive aspects of these practices not be dismissed. One notable difference between these beauty practices and modern menstrual management is that cosmetics and cosmetic surgery increased the total time and resources women spent grooming their bodies rather than replacing a previous mode of bodily care that had been experienced as burdensome. In addition, both are easily taken to extremes when women strive to realize an unrealistic, media-driven vision of a "perfect" body, one that most people will never have the resources, willpower, or biological endowment to attain.

In contrast, menstrual management resembles more closely the gender-neutral modern bodily practices of daily bathing, tooth brushing, and deodorant use, which most people can perform to their own and others' satisfaction with a moderate investment of daily time and attention but without typically inspiring ongoing anxiety or ever-greater investments of time and energy. At the same time, cosmetics and cosmetic surgery contain an important parallel to menstrual management: in a culture that increasingly saw identity as something a person performed rather than as an inherent and unchanging characteristic (Warren Susman, *Culture as History: The Transformation of American Society in the Twentieth Century* [New York: Pantheon, 1984]), these new self-fashioning practices allowed women to perform a desired class standing, even in some cases a desired race identity, linked tightly to class.

Andrea Tone's discussion of the birth control pill perhaps most closely parallels the case of menstrual management presented in this book, in that she found that women consciously accepted the sometimes-unpleasant, even potentially dangerous, side effects of birth control pills because of the tremendous benefit they experienced in terms of sexual freedom and reduction of fear of pregnancy (*Devices and Desires: A History of Contraceptives in America* [New York: Hill and Wang, 2001], 203–59). Like those interviewed for this book, the women Tone quotes could be seen from one feminist perspective as adopting a modern mode of bodily management in their lives as a substitute for making radical demands for structural change. For example, rather than using the Pill, they could have demanded social and governmental support for pregnant women, affordable daycare, legal access to abortion, and greater responsiveness of the pharmaceutical industry and research science to women's desire for safe and reliable contraception.

Likewise, with regard to menstruation, it could be imagined that women could have demanded structural changes in terms of the ability to talk about menstruation completely openly in public, at school, and at work without penalty; workplaces that accommodated menstrual-related health and comfort needs and were more generally organized with the needs of women, as much as the needs of men, in mind; greater access to true financial middle-class status through substantive reforms such as high-quality public education and a more progressive tax system; and so on. It seems clear, however, that even without making radical and conscious challenges to social structures, women were justified in feeling that their new practices were significantly better for them than what had come before, and that the net effect of these practices went a lot further toward supporting than toward undermining women's other goals, including effectively competing with men in school and at work. For another important analysis that considers women's agency in similar terms, see Judith Walzer Leavitt, *Brought to Bed: Childbearing in America, 1750 to 1950* (New York: Oxford University Press, 1986).

The major social changes of the early twentieth century are crucial to the development of the Modern Period, as demonstrated in this book. Some of those changes were initiated in, and are associated with, the "Progressive era," a historian's label that has been repeatedly deconstructed (for example, Daniel T. Rodgers, "In Search of Progressivism," *Reviews in American History* 10, no. 4 [1982]: 113–32) but that continues a hearty and productive life nonetheless (for example, Daniel Eli Burnstein, *Next to Godliness: Confronting Dirt and Despair in Progressive Era New York City* [Urbana: University of Illinois Press, 2006]). While it may not be possible to define a unitary, coherent "progressivism" characteristic of the first two decades of the twentieth century, that period did encompass a series of reform efforts of various kinds and important new ways of thinking about the individual's relationship to society and how social change ought to happen.

I highlight some of these strands as particularly important in shaping the beginnings of "modern" ways to handle menstruation, including an emphasis on personal, social, and business efficiency (see Samuel Haber, *Efficiency and Uplift: Scientific Management in the Progressive Era 1890–1920* [Chicago: University of Chicago Press, 1964]), mass public schooling, and especially programming intended to socialize new immigrants and aspiring poor and working people, including health, physical, and sex education (see Paula S. Fass, *Outside In: Minorities and the Transformation of American Education* [New York: Oxford University Press, 1989]; Jeffrey P. Moran, *Teaching Sex: The Shaping of Adolescence in the 20th Century* [Cambridge, MA: Harvard University Press, 2000]), and the socialization of many new members of a broad white- and pink-collar work force in the newly emergent modern corporation (see Olivier Zunz, *Making America Corporate, 1870–1920* [Chicago: University of Chicago Press, 1990]; Andrea Tone, *The Business of Benevolence: Industrial Paternalism in Progressive America* [Ithaca, NY: Cornell University Press, 1997]).

Other aspects of the development of "modern" American life in the twentieth century less clearly centered in the Progressive era were also important in shaping the history of menstruation. On the rise of a culture of consumption, see Gary S. Cross, *An All-Consuming Century: Why Commercialism Won in Modern America* (New York: Columbia University Press, 2000); Lizabeth Cohen, *A Consumer's Republic: The Politics of Mass Consumption in Postwar America* (New York: Knopf, 2003); William Leach, *Land of Desire: Merchants, Power, and the Rise of a New American Culture* (New York: Pantheon, 1993). On advertising, see Roland Marchand, *Advertising the American Dream: Making Way for Modernity, 1920–1940* (Berkeley and Los Angeles: University of California Press, 1985); T. J. Jackson Lears, *Fables of Abundance: A Cultural History of Advertising in America* (New York: Basic Books, 1994). On the advertising of cleanliness specifically, see T. J. Jackson Lears, "American Advertising and the Reconstruction of the Body, 1880–1930," in *Fitness in American Culture: Images of Health, Sport, and the Body, 1830–1940*, ed. Kathryn Grover (Amherst: University of Massachusetts Press, 1989); Vincent Vinikas, *Soft Soap, Hard Sell: American Hygiene in an Age of Advertisement* (Ames: Iowa State University Press, 1992).

For changes in women's work at home and in the white/pink-collar workplace, see Ruth Schwartz Cowan, *More Work for Mother: The Ironies of Household Technology from the Open Hearth to the Microwave* (New York: Basic Books, 1983); Susan Porter Benson, *Counter Cultures: Saleswomen, Managers, and Customers in American Department Stores, 1890–1940* (Urbana: University of Illinois Press, 1986).

For a compelling argument about modern reproduction and how it is different from postmodern reproduction, see Adele Clarke, "Modernity, Postmodernity and Reproductive

Processes Ca. 1890–1990 or, 'Mommy, Where Do Cyborgs Come from Anyway?'" in *The Cyborg Handbook*, ed. Chris Hables Gray (New York: Routledge, 1995). Clarke regards reproductive modernism as rooted in industrialism, and her description of its characteristics are confirmed by many of my sources, both archival and oral. I see the economic and moral valorization of industrialism (involving "standardization, efficiency, planning, specialization, professionalization, commodity and technological development, and profitability" [143]) as an important part of the modern values I describe, though not their entirety. Like Clarke, I see this kind of modernism persisting through the twentieth century, even as postmodern ideas and practices rose in importance.

The emergence of a broad self-perceived middle class is a crucial part of the development of American modernity. For a helpful overview and bibliographic essay of the history of class in the United States, which is sensitive to the importance of distinguishing between objective economic class standing and self-perceived social class standing, see Olivier Zunz, "Class," in *Encyclopedia of the United States in the Twentieth Century*, ed. Stanley I. Kutler (New York: Charles Scribner's Sons, 1996). For before 1900, see, especially, Stuart M. Blumin, *The Emergence of the Middle Class: Social Experience in the American City, 1760–1900* (Cambridge: Cambridge University Press, 1989). Evelyn Brooks Higginbotham, *Righteous Discontent: The Women's Movement in the Black Baptist Church, 1880–1920* (Cambridge, MA: Harvard University Press, 1993) describes the perspective of the African American middle class and its relationship to poorer African Americans. Marina Moskowitz, *Standard of Living: The Measure of the Middle Class in Modern America* (Baltimore: Johns Hopkins University Press, 2004) details several of the material/performance aspects of being culturally middle class.

Gallup polls are helpful in tracking the broadening middle class of the twentieth century. In 1939, Gallup conducted its first poll in which it asked people about their perception of their social class. Respondents were asked, "To what social class in this country do you feel you belong—middle class, or upper, or lower?" 86% of respondents said that they were middle, upper middle, or lower middle class. Only 5% classified themselves as lower class (Gallup Organization, Question ID: USGALLUP.39-150.QA02). Broken down fully, the responses were: upper—6%; upper middle—12%; middle—63%; lower middle—11%; lower—5%; no answer—3%. In 1952, Gallup asked the question again, this time adding the choice "working class" as well, and 36% called themselves middle class, 56% working class, and tiny minorities classified themselves as upper and lower class. By the end of the century, 30% saw themselves as working class, 62% as middle or upper-middle class, and the upper and lower class groups had shrunk almost out of existence (Gallup Organization, Question ID: USGALLUP.200036.Q42). Broken down fully, the responses were: upper—3%; upper middle—15%; middle—47%; working class—30%; lower—3%; no answer—1%.

I understand these results to indicate that by the 1930s, members of the American working class, rather than thinking of themselves as truly separate from, or in opposition to, the middle class, thought of themselves as the "working" subset of the middle class. By the end of the century, many "working" people dropped this distinction, so that the majority of Americans did not feel compelled to even qualify their "middle-class" status. Historian Steven Ross has suggested that even at the beginning of the century, well-paid manual workers frequently considered themselves to be members of the working class with reference to their workplaces, but middle class with reference to their consumption habits and their lives away from work (cited in Blumin, *The Emergence of the Middle Class*, 296; see also Olivier Zunz, *Why the American Century?* [Chicago: University of Chicago Press, 1998], 75–76). It seems that the "middle-class" aspect of many workers' identities came to predominate as consumption took on increasing importance in American culture and standards of living rose.

The most useful archival sources for studying the history of menstruation in the twentieth century include the Kimberly-Clark archives at The History Factory, Chantilly, Virginia; the online Museum of Menstruation (www.mum.org), which contains complete scanned images of, among other things, a huge number of corporate menstrual education materials that are very difficult to locate in other collections; the material culture collection at the Smithsonian, which contains examples of menstrual pads and tampons and their packaging from many decades; and Lillian Gilbreth's 1926 study of the menstrual hygiene market for Johnson and Johnson, "N" series, box 95, folder 0704-2 NGRJ3, Papers of Frank and Lillian Gilbreth, Special Collections, Purdue University Libraries, West Lafayette, Indiana.

INDEX

Page numbers in *italics* refer to illustrations.

accidents. *See* leaking blood; stains
adhesive on menstrual pads, 68, 143, 158
advertising, 10, 56, 57, 120, 121–32, *122*, 126, 139, 224n39, 230, 233; in China, 180; and false claims, 145, 150, 198; responses to, 112, 164–65; in trade press, 137, 139, 223n10
African Americans: coping with racism, 83, 153–54; demographics of interviewees, 205–6; and education, 83, 134, 225n43; extended interviewee narrative, 38–40; health beliefs, 151–53, 226n66; as Progressive reformers, 46, 234; and social class, 134, 152–57; talking to boys, 19; and tampon use, 177; as targets of Progressive reform, 46–47, 88; and work, 56, 83, 134, 225n43. *See also* slavery
agency, concept of, 5, 208, 231–32
Always brand, 158, 159, 164
amenorrhea, 23, 23–27, 33, 78–79; and pregnancy, 44. *See also* birth control pills; suppression
As One Girl to Another, 56, 88–89, 91–92, 138–39, 174. *See also* pamphlets from manufacturers

bad blood, 5, 23, 27, 29, 38, 51. *See also* plethora theory
bathing: and health, 23, 87, 90, 151–2; as marker of social class, 4, 46, 88, 128, 151–3. See also *mikveh;* odor
bathrooms: conditions of, 36, 135, 142, 150–51; daughters seeing mothers managing menstruation in, 67, 70–72; and menstrual supply storage, 31, 146, 147–48, 150, 157. *See also* toilets
belts, 30, 40, 68, 129, 185, 221n32; learning how to use, 144–45, *146*; showing through clothes, 19, 32, 125, 145
birth control pills, 101, 113, 131–32, 161, 191, 226n69; extended-cycle, 10, 197–99, 228nn4, 6, 232; and manipulation of cycles, 162–63; relieving pain, 99
blood: appearance of, 66, 98, 165; feeling of, 188–89. *See also* bad blood; menarche; plethora theory; stains; Traditional Chinese Medicine
bloodletting, 43, 214n14
Blume, Judy, *Are You There God? It's Me, Margaret,* 63, 67–69
books about menstruation. *See* health advice literature; pamphlets from manufacturers
boyfriends, 14, 65, 109, 183; in advertising, 127; assisting with menstrual supplies, 167–68; and sex, 19, 65, 109, 163
boys, learning about menstruation, 17–19, 61, 67–71, 110–11, 117
breastfeeding, 21, 25
Brumberg, Joan, 57, 134, 231
Buchan, William, *Domestic Medicine,* 21–22, 43, 79, 214n20
Buckland, Ellen, 49–50, 86–87

calendar keeping, 100, 162
Callender, Mary Pauline, 51, 52. *See also Marjorie May's Twelfth Birthday*

Canning, Kathleen, 208
carrying menstrual supplies, 31, 127, 148–49, 160, 162, 165–66; in pocketbooks, 110, 127, 149, 215n40
"catamenia," 20
cellucotton, 143–44
childbirth, 20–21, 196, 213n6, 232
children, young, 70–72, 131–32, 148, 165
chills, avoiding, 25–27, 78–79, 87–91, 90, 91. *See also* cold
China: advertising in, 180; menstrual experiences in, 32–34, 64–65, 103, 215n41, 230–31
Chinese Americans: conflicts with mothers, 103–5, 203–5, 163–64, 179–84; demographics of interviewees, 205–6, 215n41; expectations of mothers, 63–66. *See also* China; immigrants; Traditional Chinese Medicine
Christianity, 28
Clarke, Adele, 233–34
Clarke, Edward, *Sex in Education*, 75–78
class, social and economic: and advertising, 120–21, 122, 124–25, 128, 130–32; in China, 34; distinctions, 8, 31, 60–61, 88, 118, 134–36, 142, 151–54, 156–57, 170, 178–79, 224n39, 231; and establishment of broad self-perceived middle class, 6–7, 8; and gender, 4, 82–83; mobility, 4, 7, 11, 46–47, 50, 56, 88, 134, 234; nineteenth century, 22, 46, 75–76, 88; polls about, 124, 211n5, 234–35
clothing, 46, 80, 81, 127, 143, 230; concern about menstrual technology showing through, 32, 89, 125, 145. *See also* stains
cloth menstrual pads. *See* homemade pads
cold: catching, 22, 25, 50, 88–89; exposure to, 25, 26–27, 33, 78–79, 103–5, 151–52; fading concern about, 87–91, 90–91
college, 50, 57, 148, 160; in advertising, 130, 164; classes, 60; nineteenth century, 75–79; physical education at (*see* physical education); talking to friends at, 99, 108–9, 110–11, 116–17, 167, 168, 176–77; using tampons at, 185, 189–90
Comstock Laws, 44
Consumer Reports, 174
consumption, culture of, 22, 82, 125, 136, 137, 233
consumption (tuberculosis), 25, 26
contraception, oral. *See* birth control pills

corsets, 78, 79–80, 81
cosmetics, 215n40, 231–32
counterculture, 6, 72, 96, 99–100
cramps, 64, 89, 114, 115, 162, 165; discussing, 14, 63, 103, 104, 108, 109, 110, 111, 117; expert advice about, 94–96, 97, 101, 102; interfering with work, 105; relieving, 113, 160, 161, 163–64, 222n37. *See also* drugs; pain
Culpepper, Nicholas, *Compleat and Experienced Midwife*, 42–43

Dalton, Katharina, 111, 114
dating, 19, 29, 70, 99, 102. *See also* boyfriends
daughters: disagreements with mothers, 103–5, 134, 179, 181–84; experts urging mothers to educate, 21–22, 27, 45, 49, 51, 57–59, 86–87; and fathers, 108, 148; haphazard learning from mothers, 15–16, 38–39; mothers discussing menstruation with, 39–40, 42, 62–63, 70–71, 165, 187; mothers hiding menstruation from, 15, 22–23; parents using pamphlets to educate, 53, 57, 63–64; sharing menstrual supplies with mothers, 136, 188. *See also* fathers; mothers
deodorant, 121, 127–29, 154–55, 232
Depression, the, 39, 124, 134
diapers. *See* homemade pads: cloth
diet, for menstrual health, 79, 84, 101, 102, 107, 114, 115
dieting, 9, 196, 197
disgust, 19, 27, 29, 148, 187, 189
Disney. *See The Story of Menstruation*
disposability, 1, 5, 30–31, 120, 123, 132–35, 146, 160, 230; promoted in advertising, 49, 86, 121, 124, 127
disposal problems, 149–51
doctors. *See* physicians
domesticity, 22, 47, 57, 77, 131–32
douching, 127, 128, 129, 152, 174
drugs, 94–96; Advil, 104, 160–61, 164; codeine, 94–95; ibuprofen, 113, 161, 163–64, 222n37; Midol, 94, 164, 224n35
drugstores, 123, 133, 138–39, 141–42, 145, 165. *See also* purchasing menstrual products
dysmenorrhea. *See* pain

education, about menstruation, in secondary schools, 40, 41, 48, 60, 67, 68, 182, 184; for boys, 17, 110–11, 168; pre-empting parents, 61, 63–65; teaching hygiene and management, 145, 153, 156; using *The Story of Menstruation*, 53–56, 55. *See also* pamphlets from manufacturers; schools; sex education

education, demographics of interviewees, 205–6

Ellis, Havelock, 28, 217n13

e-mail, 101, 102

embarrassment: in advertising, 127, 128, 138; about buying menstrual products, 137, 139–40, 142; about forgetting a tampon, 178, 191; laughing about, 207; about menstrual blood or technology showing, 36, 69–70, 156, 189, 190; about menstruating, 15–16, 18, 20, 220n57; resisting, 60, 108; on talking about menstruation, 168

emotionality, 17, 22, 86, 93–94, 112–18, 222n43

empathy, 207

estrus, menstruation as. *See* rut

eugenics, 7, 10, 75–76, 77

exercise, 81, 86, 101, 103–4, 187, 229; to relieve menstrual problems, 78, 84, 92, 103, 107, 115. *See also* physical education; sports

experience, theorizing, 11–12, 208

fathers: hiding from, 18, 147; talking to, 108

Federal Drug Administration (FDA), 96, 197, 198, 228n4, 228n6

feminism, 49, 62, 84, 100, 101, 105–6, 108, 111, 119; in scholarship, 9–10, 57, 111, 137, 230, 231–32

flooding, 24, 42

"flowers, the," 20

Foote, Edward, *Plain Home Talk*, 26, 45

Foucault, Michel, 10, 212n8

Freethinkers, 43–44, 45

friends: in advertising, 127, 129, 130; boys as, 17–18, 69, 97, 168; girls, as source of information, 14–15, 16, 39, 40–41, 58, 61–62, 68, 102–3, 104, 177–78, 180, 182, 183, 188, 190; and teasing, 110, 117, 153

fright, 23, 25, 27, 38–39, 40–41, 57, 89, 213n8

generational change, 6, 23–24, 29, 67, 72, 79, 89, 101, 105, 159, 187, 188, 226n66; interviewees reflecting upon, 16–17, 34, 40, 64, 153, 166

Gilbreth, Lillian, 143–44, 150, 235

grandmothers, as source of information, 59, 60, 89, 104

Greek medicine, ancient, 24, 27

Green, Monica, 20, 229

guilt felt by mothers, 58, 63

Gunn, John, *Domestic Medicine*, 26, 43

health advice literature, 21–22, 42–46, 47–48, 65, 97, 101–2, 176. *See also* pamphlets from manufacturers

health beliefs: general, 1, 5, 23–29, 33, 75–79, 89–92, 100–101, 103–5, 134; reproductive, 26, 75–76, 106. *See also* cold; health advice literature; industrial hygiene; pamphlets from manufacturers; physical education; tampons, and safety concerns; Traditional Chinese Medicine

heat, menstruation as. *See* rut

heroic medicine, 43

Hollick, Frederick, 26, 45

Hollingsworth, Leta Stetter, 82

homemade pads: cloth, 30–32, 34, 35, 38–39, 87, 92, 134–36; cotton, 35–36; and environmentalism, 160; paper, 33–34

humoral theory, 24–25, 89–91, 214n14

husbands, communicating with, 18, 213n6; about buying menstrual products, 141–42; about mood, 94, 115–16; about sex, 19–20, 29, 33, 98

hygiene: broadly defined, 79–80, 81, 84–85, 93; as cleanliness, 46, 49, 146, 170. *See also* industrial hygiene

hymen, and tampon use, 173–76, 179

immigrants, 35, 40, 134; from China, 33–34, 65, 103–5, 179–84, 205–6, 215n41; nineteenth century, 10, 76; Progressive era, 46, 88, 134, 224n39, 233. *See also* Chinese Americans

impurity, 24, 38, 51–52

industrial hygiene, 83–86

industrialism, 2, 121, 231, 234

Instead cups, 160, 164

interviews as historical source, 11–12, 205–10

irregularity, 25, 42–43, 68, 74, 161

IUD (intra-uterine device), 67

Jacobi, Mary Putnam, 77–78, 79

Japanese internment during WWII, 35–36

Jewish community, 15, 23, 40
Johnson & Johnson, 52–53, 143, 150, 235. *See also* Modess brand
Jones, Marion, 56
Journal of the American Medical Association (JAMA), 170, 174

Kellogg, John Harvey, 45, 78, 79
Kimberly-Clark: archives and analysis, 230, 235; business decisions about advertising, 123, 125, 145; business decisions about retailing, 136, 137–41; business decisions about tampons, 171, 175; education department, 48–52, 53–57, 59, 86–89, 174–75; market share, 136; product engineering, 143–44, 154, 172. *See also* Kotex; pamphlets from manufacturers; *The Story of Menstruation*
Kotex: altering before use, 5, 144; brand preference, 158–59, 164, 191; as generic name for sanitary napkins, 136, 142–43; product design, 49, 143–44. *See also* advertising; disposability; disposal problems; Kimberly-Clark; purchasing menstrual products

Ladies' Home Journal, 56, 101, 122, 123, 126, 184, 223n8
Ladies' Physiological Societies, 44
laughter, 206–7
leaking blood, 100–101, 120, 143–44, 164, 172; fear of, 186; teasing about, 18. *See also* embarrassment; stains
Lears, T. J. Jackson, 121, 223n20, 224n39, 233
libraries, 40–41, 48, 60, 63
Lybrel, 228n6
Lyman, Henry, 78

magazines: advertising, 19, 57, 123–24, 223n10; articles, 60, 69–70, 101–2, 112, 114, 176. *See also Ladies' Home Journal*
Malchow, Charles, 27–28
manners, middle-class, 46
Marjorie May's Twelfth Birthday, 51, 56, 57, 87, 92. *See also* pamphlets from manufacturers
marriage, 22, 47, 50, 55, 77, 94; and becoming sexually active, 96, 177, 178; and sex, 27, 29. *See also* husbands
Martin, Emily, 219n55, 228n1, 231

mass market, 81, 120, 132, 136, 157
men and boys, talking with, 71, 106, 108–11, 166–68. *See also* boyfriends; boys learning about menstruation; fathers; husbands
menarche, 10, 21–22, 51, 53, 57, 148, 158, 231; neutral or positive experience of, 39–40, 158; as surprise, 14, 38–39, 40, 64, 213n8
men menstruating, 24
menorrhagia. *See* flooding
Metropolitan Life Insurance, 84–85, 142
Middle Ages: in China, 33, 230; in Europe, 20–21, 24, 25, 27, 229
mikveh, 15
modernity: desire for, 5; popular conception of, 3–5, 233; relationship to social and economic class, 4; reproductive, 233–34
Modess brand, 52–53, 93, 124, 125, 126, 136, 139, 143, 227n15. *See also* advertising; Johnson & Johnson
"monthlies," 20
morality, 22, 27, 29, 76, 129, 234
Mosher, Clelia Duel, 81–82, 84
mothers: in advertising, 125, 126, 132, 224n38; advice to, 21–22, 27; communicating with daughters about menarche, 10, 14, 57–59, 62–64; communicating with daughters about menstrual management, 30, 35–36, 67, 70–72, 102, 136, 158, 165, 166, 187–88; communicating with daughters about sex, 16, 40, 58–59; communicating with sons, 168; conflicts with daughters, 103–4, 163–64, 179, 181–82; expectations of, 63; hiding from, 15, 112, 134, 182, 185. *See also* daughters; sons

natural, menstruation as, 16–17, 49, 50, 52, 53, 72, 86, 161, 163, 164, 229–30
new women, 47, 49–50, 75–76, 131; as flappers, 50, 125, 130
Northcote, Hugh, 28
Novak, Emil, 23

odor, 19, 31, 88, 128–29, 148, 152–55, 171, 178, 189, 191, 223n20. *See also* deodorant
openness, ideal of, 6, 10–11, 42, 58, 69, 70, 72, 141, 183
ovulation, in relation to menstrual cycle, 26, 44–45, 51–52, 53, 65, 217nn11–13

packaging of menstrual products, 136–41, 142, 147, 174, 181–82, 235
pain, 25, 50, 64, 78, 81–82, 112, 115; addressing, 66, 74, 80, 83–87, 94–96, 161; caused by chafing, 34, 92; caused by tampon, 178, 186; impeding activities, 101, 105, 119. *See also* cramps; drugs
pamphlets from manufacturers, 39–40, 61–62, 138–39, *146*; explaining menstrual cycle, 48–59, *54*, 67; giving health advice, 86–89, 90–91, 92, 93, 174–75; promoting products, 49; teaching how to use menstrual products, *146*
parenting, 38, 40, 44, 49, 53, 63, 71, 182–83. *See also* daughters; fathers; mothers; sons
patent medicines, 43
personality, 93, 129–30, 224n39
physical education, 14, 48, 79–82, 92, 103–4, 231. *See also* exercise; sports
physicians, 10, 20–21, 22, 24–27, 76, 152, 163; addressing pain, 83–84, 94–96; and inadequate treatment, 106–7; teaching about the body, 39, 42. *See also* industrial hygiene; tampons: and safety
Pinkham, Lydia E., 43
plethora theory, 24–27, 42–43, 44, 45–46, 89–91
PMS. *See* premenstrual syndrome
PMSing, as slang term, 102, 116–19. *See also* premenstrual syndrome
pocketbooks. *See* carrying menstrual supplies
pregnancy, 20–21, 28, 41, 67, 161, 162, 232; boyfriends' concern about, 167–68; mothers' concerns about, 16, 40, 136, 179; as possible source of amenorrhea, 23, 24, 25, 26, 44
premenstrual syndrome (PMS), 101, 102, 106–7, 111–18, 164, 222n43
privacy, lack of, 36, 146–47, 148–49, 190
privies, 31
product design, 69, 125, 158–59. *See also* Kotex, product design
Progressive era, 2–3, 5–6, 46–48, 88, 134, 192, 232–33
prostitution, 10, 28
puberty, 22, 26, 61, 75, 78, 229. *See also* menarche
public, menstruation in, 116, 120, 123, 140, 141, 151, 169, 208, 232
purchasing menstrual products, 11, 68, 136–41, 143, 158, 166, 185; by husbands and boyfriends, 141–42, 166–67; from vending machines, 142, 148, 151

"rag, on the," 17, 19, 112, 117, 118, 135, 145
rags. *See* homemade pads, cloth
reproduction, 12, 20–21, 75, 208–9, 233–34; learning about, 21, 22, 39, 43–44, 53, 60, 65; and menstrual cycle, 24, 41, 44, 67, 77, 106. *See also* ovulation
respectability, in middle-class culture, 47, 88, 157
rut, menstruation as, 26, 27, 45

schools, secondary, 6, 10, 75, 77, 233; first period at, 39, 59; friends as confidantes at, 16, 168; friends sharing information at, 14–15, 17, 58, 135, 191; hiding menstrual supplies at, 31, 148–49; nurses' offices, 64; physical education in, 14, 92, 103–4. *See also* college; education about menstruation in secondary schools; teasing
science, popular, 2, 42–46, 49, 121
scientific explanation of menstruation for children, 38, 40–42, 51–55, 65–66, 67. *See also* education, about menstruation, in secondary schools; sex education
scientific research on menstruation, 44–45, 51–52, 77–78, 82, 85–86; and tampons, 171–74, 175
Scott, Joan, 208
Seasonale. *See* birth control pills
"Secrets of Women" literature, 20, 24
self education, 38–41, 58, 60, 62, 63, 64, 185. *See also* health advice literature
sex during menstruation: and coercion, 29; conflict over, 109; healthfulness of, 27–28, 96; and hygiene, 97–98; and morality, 29; and sexual revolution, 99–100
sex education, 52, 57, 61, 63–64, 67, 96–97, 164, 176; and boys, 17, 61, 110; Progressive era, 47–48, 233
sexual desire, 27–28, 173, 217n13, 226n10
shame, 7, 13, 16, 20, 33, 38, 40, 72, 136, 148, 153. *See also* embarrassment
shock: mental, 14, 25–26, 79, 89–91, 213n8; physical, 25–26, 88. *See also* cold; exercise
sisters: as confidants, 18, 147; not sharing

sisters (continued)
  information, 16, 38–39; as source of information, 15, 16, 23, 41–42, 64, 187–88; talking to brothers, 108–9, 117
slavery, 22–23
Smith College, 81
sons: hiding from, 18–19; talking with, 17, 18–19, 65–66, 71, 168
sports: expert advice about, 95, *126*, 224n34; interviewees playing, 14, 92, 100–101, 189; and Title IX, 62. See also exercise; physical education
stains, 98, 144, 155–57, 159, 162. See also leaking blood
Stockham, Alice Bunker, 78, 79
Story of Menstruation, The, 53–56, *54*, 59
Strasser, Susan, 30, 144, 230
suppression, 23, 25–27, 33, 42, 79, 89–91. See also amenorrhea
swimming, 23–24, 81, 88–89, 100, 103, 152, 189; and tampons, 125, 130, 178, 180, 187. See also cold; exercise; sports

taboo, menstrual, 20, 27, 28, 45, 140, 148, 156, 217n13, 229–30
Tampax, 58, 81, 120, 123–24, 159, 172, 174, 184, 226n68. See also advertising; tampons
tampons: advertising, 123–24, 131; applicators for, 226n68; and bathing, 128; children seeing, 67, 71, 148, 166; efficacy, 92–93, 125, 129, 159, 160–61, 172, 186; failure to remove, 178, 186–87, 191–92; learning to use, 164, 184–87; medical use, 5, 170; o.b. brand, 159, 226n68; and peer pressure, 189–90; and safety concerns, 104, 170–72, 181–82, 189, 190–92; sexual implications, 173–76, 177, 179, 182–84; and social identity, 8; storing, 70, 127, 147, 148. See also swimming: and tampons; Tampax; Toxic Shock Syndrome
teasing, 118, 148; by boys, 19, 110, 145, 149, 155–56; by girls, 152–53

television, 19, 69, 112, 158–59, 164–65, 198; in China, 180
toilets, 17, 18, 71, 127, 146, 149–50
Toxic Shock Syndrome (TSS), 102, 181–82, 183, 186, 187, 189, 190–92; effects on tampon market, 227n22
Traditional Chinese Medicine, 33, 104–5, 163, 181
Trotula, 20

underwear, 14, 68, 142, 159, 165
uterus, 28, 78, 80–81, 95, 171; in eighteenth- and nineteenth-century theories of menstrual cycle, 25–26, 45; in scientific explanations for girls, 41, 50, 51–52, 53, 67

Verbrugge, Martha, 80, 217n9, 231
vernacular tradition, 22–23, 214n11, 229
Very Personally Yours, 53–54, *54*, 56, 57, 175
vicarious menstruation, 26
Victorian era, 22, 27, 47, 75–76, 103, 229
virginity. *See* hymen

washing cloth pads, 30–31, 160
Wings, 159, 165
womb. *See* uterus
women's community, and reproductive labor, 20–21, 213n6
women's history, 9–10, 229–34
Woodward, Elizabeth, 56, 219n46
work, 2, 56, 230, 232, 233; buying menstrual products at, 142; concealing menstruation at, 160; incapacity to, 25, 94–95, 105, 108, 111; industrial, 46, 53, 76–77, 131, 133, 224n35; reproductive, 20–21; talking about menstruation at, 110, 116, 165–66, 180–81; white- and pink-collar, 1, 4, 82–84, 121, 130–31, 224n34. *See also* industrial hygiene
World War II, 35–36, 53, 131